Dr Ronald Ross

mosquito, malaria, India and the Nobel Prize –
an untold story of the First Indian Nobel Laureate

About the Authors

Born 23rd July, 1951, **Prof. Dr B.K. Tyagi** completed MSc (Zoology) from Meerut Univ. in 1974 and PhD (Entomology/Zoology) from Garhwal Univ. in 1978. He emeritated in 2013, with two years' extension till 2015 post-emeritation, as Scientist 'G' (Director) & Director in-Charge from the ICMR's Centre for Research in Medical Entomology, Madurai after 35 years of long and highly distinguished career in medico-entomology researches in most States and UTs in the country, winning in course the ICMR's prestigious Dr MOT Iyengar Memorial Award (2008) and the TDR-WHO's First Prize in a worldwide competition (1995). Prof. Tyagi has for past several months been working as the Advisor, VIT Vellore (TN, India).

Prof. Tyagi has to his credit >610 scientific titles, plus 35 books, incl. a WHO Training Manual (World Health Organization, pp. 240; Yr. 2015). He holds a patent (India Patent # 191635). He has served as the First National Representative in-Charge, National Office in India in 1981 onwards (later since 1988 also as the Chairman, SIO Regl. Off. in Southern Asia) for the International Odonatological Society and the Odonata Specialist Group of the Species Survival Commission (IUCN) in 1985 onwards. Currently he is functioning either as editor or member of Editorial Board for >3 dozen international/national journals. In addition, he is also serving as an expert on various research committees of ICMR, DRDO, DST, DBT-RCGM, MoE&FCC etc. and is Chairman, RCGM/DBT Sub-Committee on GM *Aedes aegypti*. He is the present founding President of Society of Medical Arthropodology and Indian Dragonfly Society.

Dr. Sajal Bhattacharya, presently Associate Professor at the Post Graduate Department of Zoology, Asutosh College, India, was born and brought up in Kolkata.

Dr. Bhattacharya is an alumnus of University of Calcutta and has carried out his doctoral research at the Department of Medical Entomology, Calcutta School of Tropical Medicine. He pursued his post-doctoral assignment at IHMT, Lisbon. His field of research is disease vector eco-epidemiology and he has more than a hundred scientific publications to his name. He is an elected Fellow of the Royal Society of Tropical Medicine and Hygiene, London and has been invited to various countries, namelyAustralia, Hungary, Italy, England, Kenya, Indonesia to present his work. He had been invited to participate in a WHO (Africa region) organized meeting, in 2017, at the Republic of Congo. In the following year he was honored by the Government of West Bengal with the Outstanding teacher award. He has also published three papers on the life and works of Sir Ronald Ross. He was conferred the prestigious Lifetime achievement award by the Society of Medical Arthropodology (SOMA) at Sir Ronald Ross Institute of Parasitology, Hyderabad, India.

Prof. B. Reddya Naik is Professor of Zoology, Osmania University, Hyderabad. He joined university services as Assistant Professor on 2nd Sept. 1997 and received a Post-Doctoral Research Award from UGC Selection and Award Bureau New Delhi (2004-2007). His broad area of research is Medical Entomology with special emphasis on mosquitoes and mosquito-borne disease surveillance and their preventive measures by using larivorous fish and plant products. He has published 36 research papers in peer reviewed National /International journals. He has guided 5 research scholars for their Ph. D degree and is currently guiding another 8 research scholars. He has attended /participated /presented research findings in 38 National/International seminars /Conferences /workshops and organized 27 Seminars/ Symposium /Workshops /Conference /Training Programmes in the respective professional/ research field. Resource person to Academic Staff College, besides producing a documentary film on "Chikungunya" in Collaboration with EFLU, EMMRC, Hyderaba. He has visited USA in 2008 and China in 2013 for presentations research papers in the seminar /workshop. Other than teaching for the past 22 years, he has rendered services to University in different capacities; as Director to Sir Ronald Ross Institute of Parasitology, Begumpet from May 2008 to Feb 2012 and given a shape to the Nobel laureate's work place and brought the institute into lime light. He has conducted many courses as the NSS Programme coordinator of Osmania University from July 2014 to Nov.2015 and was Local Secretary for the 105th Indian Science Congress - 2018.

Dr Ronald Ross

mosquito, malaria, India and the Nobel Prize

an untold story of the First Indian Nobel Laureate

Dr B.K. Tyagi

Emtd. Scientist 'G' (Director) & Director in-Charge
ICMR - Centre for Research in Medical Entomology Madurai, TN, India;
Advisor, VIT University, Vellore, Tamil Nadu, India;
Consultant, BCIL-UN FAO Sri Lanka Project on Biosafety
&
President, Society of Medical Arthropodology

Dr Sajal Bhattacharya

Associate Professor & HOD,
Department of Zoology,
Asutosh College,
Kolkata, West Bengal, India

&

Dr B. Reddya Naik

Professor,
Department of Zoology,
Osmania University,
Hyderabad, Telangana, India

Society of Medical Arthropodology
Jodhpur, India
2020

Published by

SCIENTIFIC PUBLISHERS (INDIA)
5 A, New Pali Road, P.O. Box 91
Jodhpur 342 001 (India)
E-mail: info@scientificpub.com
Website: www.scientificpub.com

ISBN : 978-93-89412-39-0
eISBN : 978-93-89412-40-6

Printed in India

Dedicated to

all those scientists who have given
their everything – time, energy and resources,
for a better understanding of the man's deadliest foes
in biological history
– mosquito and malaria –
for a happier world to live in.

Foreword

Dr Col Sir Ronald Ross won the 1902 Nobel Prize in Medicine or Physiology for his discovery that Malaria was caused by the mosquito Anopheles and not "mal' air" as it was widely believed. It is said that there are several thousand writings about him but this book is unique in that it is written completely with an Indian perspective by three eminent Indians hailing from different parts of this country. It is aptly titled "Dr Ronald Ross: Mosquito Malaria India and the Nobel Prize – an untold story of the First Indian Nobel Laureate."

It is my proud privilege to write a foreword for this particular authored book because I hail from the same region where the discovery was made. I am really proud to be an Indian Army Officer born and brought up in Secunderabad in close proximity to the place where the epic discovery was made by a serving Major of the Indian Medical Services (Royal British Medical Corps). It is a strange coincidence that around this place of discovery there are very old learning centres of the same period or slightly prior to it. These are namely, The Jagirdhar College now called The Hyderabad Public School, the first Post Grad College at Secunderabad and Mahbub and Wesley Colleges. Incidentally the learning centre for the erstwhile Satyam Computers was also situated here. Very few of us know that Swami Vivekandandaji visited Mahbub College before leaving for Chicago two years before the discovery period. This is the zone of influence of the place.

Another point to note is that Late Sri Rajeev Gandhi was born on the 20th of August which happens to be the date of the discovery and was trained in this same building as a pilot officer. Strangely this building was constructed in brick with mud mortar and lime plaster as it was meant for the functional requirement of a 20 bedded hospital for Indian troops. At this humble place Sir Ronald Ross worked for his discovery using his spare time and utilizing his own funds for the research. His superior officer was not favourable to this work as it was not in the charter of duties and he was not available for the club and play activities. With all these constraints he had worked on stoically to achieve this discovery. He had become a medical doctor much against his own leanings as his father wished him to be one. His father who was a serving senior military officer here also encouraged him to work for a solution for the malaria fever that was causing so much devastation in the Indian subcontinent. His devotion to this cause made him live within the campus with his family to be at walking distance, forgoing the bungalow type accommodation provided for him in the Secunderabad Cantonment with all luxuries befitting his status. We can note with interest that after dissecting several numbers of mosquitoes it was from the 20th one that he finally made this discovery.

Thus the significance of this momentous finding of Dr Ronald Ross should be given its due regard because he had faced many tough and hostile professional situations in the I.M.S. Knighted and honoured with a large number of distinctions all over the world, the achievement closest to his heart was always the Nobel Prize won for his mosquito-malaria research in India. He is, therefore, truly the first Nobel Laureate of India!

I as an Army Officer have been associated with this building of discovery for the last three decades and inspired several students of medicine, architecture and schools to work for a cause for humanity by organizing innumerable visits to this august place. Any visitor coming to Hyderabad has been encouraged by me to visit this place without fail. The batch of students from Karaikal near Pondicherry who were affected by the Tsunami of 2004 was enthralled by a visit. And the birthday of one of these children happens to fall on the birthday of Sir Ronald Ross which made her declare that she too will serve India in no less stature by inventing an early warning system to detect tsunamis and other such calamities. Such is the pull factor of this place!

Written in a simple and clear language, this book by Professors B.K. Tyagi, Sajal Bhattacharya and B. Reddya Naik, is certainly a notable and unique biography. In it the hitherto lesser known aspects of the life of this great person, Dr Col Sir Ronald Ross, particularly during the period of his life spent in India, are highlighted. I am certain that the readers will find it most interesting and empowering to follow their dreams with courage, conviction and commitment which is our Indian heritage.

Col N.A. Kumar
Secunderabad

9th September, 2019

Preface

"Nearly all the ideas in science are provided by amateurs."

– Dr. Ronald Ross (1923)

(Winner of The Nobel Prize in Physiology or Medicine, 1902 – for mosquito-malaria research in India)

A large number of vector-borne diseases are a reason for over a million mortality and several millions morbidity throughout the world every year. Among all these, for centuries, malaria had been devastating peoples' health, economy and intelligentsia in many countries mostly in tropical and subtropical regions. In India the disease has been known since antiquity and has been a major cause for her poverty, low agriculture productivity, high mortality particularly among children below 5 years of age, decreased knowledge quotient and general public health. The early Indian and Greek literature of yore had well documented the characteristic poor health due to malarial fevers and enlarged spleens. For over 2500 years the idea that malaria fevers were caused by miasmas rising from swamps persisted and it is widely held that the word malaria comes from the Italian mal'aria meaning polluted air although this has been disputed. With the discovery of bacteria by Antoni van Leeuwenhoek in 1676, and the incrimination of microorganisms as causes of infectious diseases and the development of the germ theory of infection by Louis Pasteur and Robert Koch in 1878-1879, the search for the cause of malaria intensified. Scientific studies only became possible after the discovery of the parasites themselves by Charles Louis Alphonse Laveran in 1880 and the incrimination of mosquitoes as the vectors, then first for avian malaria by Ronald Ross in 1898 and then for human malaria by the Italian scientist Giovanni Battista Grassi and his team including Amico Bignami, Giuseppe Bastianelli, Angelo Celli, Camillo Golgi and Ettore Marchiafava in late 1898.

Road to success is often full of thorns; its as difficult as to tread on a double-edged sword. Around the same time of the above mentioned pathbreaking medical discoveries, announcement of setting up of the Nobel Prize in Medicine and Physiology (as a part of a series of world prizes instituted in 1896 following the death will of the dynamite king Alfred Nobel, a Swede who had a great benevolent heart and appreciation for scientists' remarkable contribution in making a better world to live in), with the largest monetary payments ever for scientific discoveries, apparently ignited every mind around genuinely engaged in deciphering malaria-enigma and to proclaim first to reach the ever awaited conclusion! The contest eventually narrowed down to two scientists of contrasting characters, namely Giovanni Battista Grassi, a zoologist of international repute who had been seriously toiling to prove a malaria-mosquito connection for several years, and Ronald Ross, a exceptionally energetic, hardworking and self-trained scientist who was pursuing his malaria research India under the guidance of Dr Patrick Manson, in England, but still could not distinguish one mosquito from another. Once good friends, they turned out to be deadly foes to each other's reputation on the issues like primacy and priority of their research on mosquito-malaria connection and parasite life cycle outside the human body, in the mosquito body. The international jury had to settle the matter in favour of Dr Ross who was awarded the Nobel Prize in 1902 (actually handed out in 1904 only). For whatever happened between these two geniuses the humanity was greatly benefited from their tireless inquiries in the mystique mosquito-malaria relationship and what benefits we reap today in saving precious lives to the savage attacks of the insidious malarial disease is due entirely to their, and other similarly devoted scientists', tireless and focused hard work.

Ronald Ross was born in Almora located in the Himalaya in northern India on 13th May, 1857. He completed his schooling as well as medical education in England where after he had joined Indian Military Service in 1881. A medical doctor who received the Nobel Prize for Physiology or Medicine in 1902 for his work on the transmission of malaria, becoming the first Indian Nobel laureate, and the first British born outside England or Europe. His discovery of the malarial parasite in the gastrointestinal tract of a mosquito in 1897 proved that malaria was transmitted by mosquitoes, and laid the foundation for the method of combating the disease. He was a polymath, writing a number of poems, published several novels, and composed songs. He was also an amateur artist and natural mathematician. He worked in the Indian Medical Service for 18 years. It was during his service in Secunderabad, India, that he made the ground-breaking medical discovery on 20th August, 1897. After resigning from his service in India in February, 1899, he joined the faculty of Liverpool School of Tropical Medicine, and continued as Professor and Chairman of Tropical Medicine of the institute for 10 years. In 1926 he became Director-in-Chief of the Ross Institute and Hospital for Tropical Diseases, established in his honour, and remained there until his death on 16th September, 1932. While Ross will be principally remembered for his malaria work, this remarkable man was also a mathematician, epidemiologist, sanitarian, editor, novelist, dramatist, poet, amateur musician, composer, and artist. He was truly a genius.

But Dr Ross won this great feat in his highly tumultuous and checkered career with the help and assistance of many associates. Save for Dr Manson and those in the I.M.S., he nearly never referred to anyone of the native helps, particularly Kishori Mohan Bandhyopadhyay who was the man anchor architect for all his laboratory proceedings in malaria research on birds in Calcutta, and it is understood by the ensuing awards to Kishori Mohan that the British Empire had recognized a major mistake in hiding his contribution toward Ross's epochal discovery and compensated him instead in the manner as stated above.

In spite of all the minimum helps and opportunities, Dr Ronald Ross proved to be one of those rare scientists who by their energetic character not only practically educate the budding researchers in habits of industry, but by the example of diligence and perseverance which they set before them, largely influence the scientific activity in all directions and contribute in a great degree to form the national character. The national progress, it is aptly said, is after all the sum of individual industry, energy and uprightness. His was unquestionably a mind of the very highest order. He was a thinking scientist – a genius.

Dr Sir Ronald Ross KCB, KCMG, FRS, FRCS (1857–1932), best known for his pathbreaking medical discovery of 'mosquito-malaria connection' and for elucidating the malaria parasite's life cycle outside the human body, for which he was awarded the Nobel Prize in Physiology or Medicine, in 1902, is one of the most celebrated scientists in medical history who had been written about exceptionally exceedingly during past nearly eight decades; whoppingly more than 30,000 titles are devoted to him! For a humongous subject like writing another book about his life and works, it needed a Himalayan determination to add further new information with a different flavour to his already much illustrated career. To achieve this motto one of us (Prof. Dr B.K. Tyagi) had been researching on his life and works for last nearly two decades since 1999, collecting all possible literature about Dr Ross. The theme of the current book is, therefore, entirely different from the rest of biographies as we have throughout highlighted the Indian side of his discoveries, research and career.

"Dr Ronald Ross: mosquito, malaria, India and the Nobel Prize – *an untold story of the First Indian Nobel Laureate*" is a unique book, incorporating fables unrecounted so far, and written in simple and lucid language. The Book – by Professors B.K. Tyagi, Sajal Bhattacharya and B. Reddya Naik, is biographically presented with the unknown or less known aspects of life of this great malariologist, Dr Ronald Ross, particularly the part spent in India. These have been highlighted to bring on to fore a trail of reinvigorating gust of fresh information and motivation to young and budding researchers in India, in particular, to learn from the Nature and never give up under any circumstance, and that there is no alternative to hard work and focused application.

B.K. Tyagi

20th August, 2019 **Sajal Bhattacharya** and **B. Reddya Naik**

Acknowledgement

It is unimaginable to encompass in a small composition like this a vast and diverse life attribution dedicated single-mindedly to the cause of discovering the role of mosquitoes in malaria transmission without the immense help of the generous Librarians of various national institutional libraries particularly those of National Centre for Diseases Control, Delhi, Vector Control Research Centre, Puducherry, National Institute of Malaria Research, Delhi, Regional Medical Research Centre, Dibrugarh (Assam), Indian Institute of Science, Bengaluru, Osmania University, Telangana and Hyderabad University, Hyderabad etc.

Discussions with and various materials, including a compact disc, received from Col. (Retd.) A. Kumar, Indian Army have been particularly helpful in accomplishing the task. Our heartfelt thanks go to Col. (Retd.) A. Kumar for also contributing a highly delightful and inspirational 'Foreword' for this book.

We thank some of our scholars, students and other informants for their genuine help with the various literary materials related to the life and works of Dr Ross, such as Dr Sambashiva Daravath, Assistant Professor & I/C Head, Dept of Biotechnology, Nizam College (OU), Basheerbhag, Hyderabad; Mr Siddaiah Madpathi, Research Scholar, Department of Zoology, UCS Osmania University, Hyderabad; Ms Swaha Bhattacharya, San Jose, California, USA and Dr. A.K. Mandal, and Ms Chandrima Bose, both from Post-Graduate Department of Zoology, Asutosh College, Kolkata as well as the Director, I.P.G.M.E.&R. (SSKM Hospital Campus), Kolkata, India.

As always we invariably thank our better-half: Ajita (B.K. Tyagi), Sanchita (Sajal Bhattacharya) and Kamala (B. Reddya Naik), who were all supportive and motivating to us throughout the period of completing this rare composition.

Last, but not least, we owe much to Mr Pawan Kumar Sharma, CEO of the Scientific Publishers (India), Jodhpur, for his trust in us and for the patience and help extended throughout the process of writing this monograph.

Contents

PART I: ORIENTATION

1. Family and early childhood - ***Like begets like*** (1857-1865)

1

Family and Early Childhood - *Like Begets Like* (1857-1865)

"The child is the father of the man"

– (William Wordsworth)

In the life of Dr Ronald Ross, early childhood environment had played a very important role. The family legacy of valour during military actions on frontier areas, parental solicitude in the midst of pristine lap of Himalayan township of Almora, and closeness to local Indian folks had all left on his persona an indelible imprint for his future character building of an inexhaustible and energetic scientist. It is considered opportune, therefore, to sketch here foremost, with a tinge of dramatization, the environment then prevailed; on one hand, an eye of the storm in the form of 'Sipoy Mutiny' and bad omen associated with his date of birth, and, on the other hand, the cool and quiet environment of high Himalayan mountain military station, Almora - his place of birth, for inveigling a perfect melange of diverse knowledge - a state of polymath!

1857 – arguably the most important year in the life of India then struggling for her freedom from nearly one hundred years of brutal rule by the British Empire (Fig. 1.1). The Great Indian Mutiny had just emerged on 10th May in Meerut, a major agriculture town in the western part of United Provinces (now Uttar Pradesh) used for settlement of massive military cantonment here and many other places in the neighbouring States, and the British Government in India was on way to announce the infamous 'Siege of Delhi' later the same year. Heat of the deafening roaring of "Har Har Mahadev" and "Jai Hind" of the millions of Indians, both civilians and the armed constabulary in the British India, could be sensed all over the country. The mutiny by the Indian soldiers had beguiled *de novo* a new beginning in the fight against the tyranny of a century old foreign rule which had not before much long perceived its obligatory exit from India. The real battle, a mélange of both non-violent and violent tribulations, had just begun! As if by providence, ironically though, almost at the same time seeds for another kind of fierce battle - this time a biological one between man and malaria, through the latter's carrier, the mosquito, had been sown which would later mark the beginning of a new era of research with the disease vector appearing in a pivotal role!

The generally quiet environs of the mighty Kumaun Himalaya in northern United Provinces (now Uttara Khand) was also put on high alert owing to the rise of the countrywide 'Sipoy mutiny'. The British Army regiment in Almora, a serene and sparsely populated hilly habitat in higher ranges of the mountain, was suddenly awakened with a streak of blitzkrieg military maneuvers prepared to suppress and keep the impact of mutiny at bay. The high mountain wind blew through lofty alpine jungle surrounding the cantonment area with a soaring thunder indicating for a possible storm. Campbell Clave Grant Ross, a young major in the British Indian Army was a tough task master for his men in uniform, with terse face and hardly any expression. In his oversized office tucked under the beautifully trimmed trees, he could be seen trolling incessantly in a pensive mood. He was in the first place thinking about his preparedness for meeting against the mutiny's towering inferno, but there was another thought as well that continuously occupied his mind. His wife was pregnant and they expected the baby anytime then!

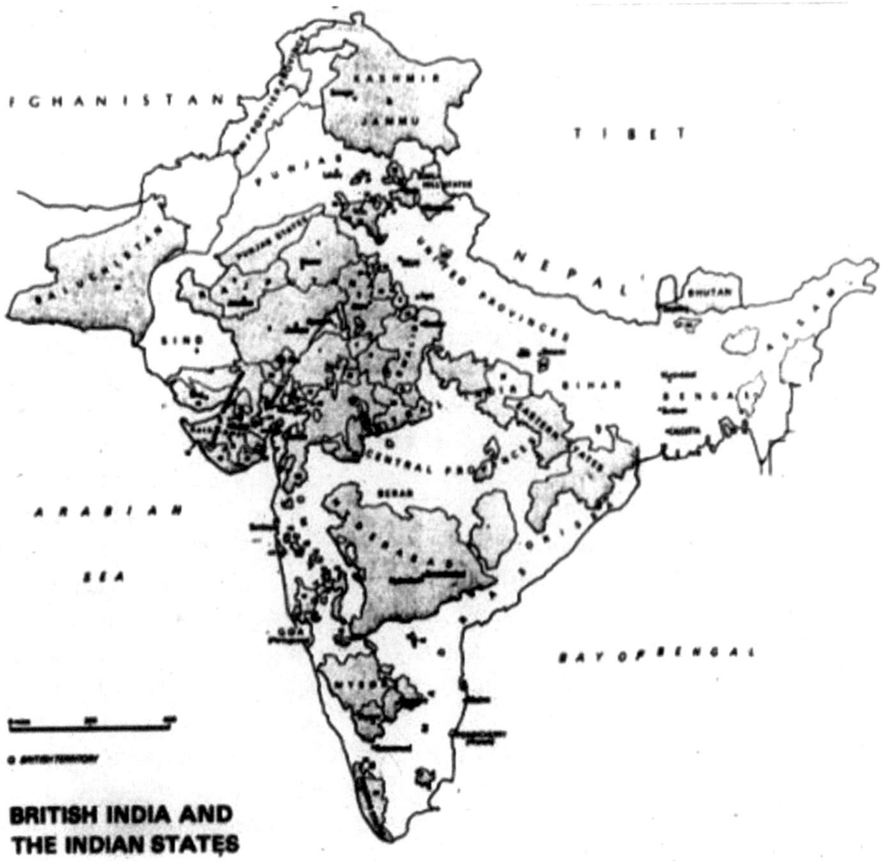

Fig. 1.1. British India and the Indian States (*Source:* courtesy Col. A. Kumar's PPT slide)

Outside his sprawling office the wind was only intensifying and many makeshift accommodations, *Chowldari,* for soldiers and other supporting staff, were frill-flapping with frightening echo and threatened to be uprooted anytime. Major Ross was pensive for several different compelling thoughts, but the one that occupied most of his attention was the thought of arrival of his first child. He grew impatient with every passing moment and unmindfully walked the length and width of his room. Just then, he heard a knock at the door. He gave permission to come in whereupon a sentry showed up to inform the news of the birth of a baby boy, and both the child and mother were in good health. He was immensely elated and quickly moved to a side table to collect his pea cap and the baton. Just while leaving he chanced upon looking at the calendar on his working table and gazed unblinkingly for many seconds; the date was 13th May, the superstitiously perceived an unlucky day. Against all odds, however, the world would decades later acknowledge the genius of this man, Ronald Ross, as the discoverer of relationship of the mosquito with malaria transmission in man – the most intriguing question on Earth then, which was certainly absolutely inevitably the starting point of Ariadne's thread that would eventually lead to the exit from the labyrinth of malaria and help mankind get rid of the deadliest scourge one day. He went on to win the Nobel Prize for Medicine in 1902 for his path-breaking discovery!

Ronald Ross was born in Almora, a cantonment town in the United Provinces (now Uttara Khand), located at a height of about 1650 meters (5500 feet) at the southern end of the Kumaon Hills in the Himalayan region, on 13th May, 1857. In those days Almora was a hill resort best known for, besides a rather small village on the foothill, the massive military cantonment and certain other settlements the British had founded to facilitate their establishment such as massive playgrounds (Fig. 1.2). The picturesque hill station is ringed by mountains, forests of fir and pines and the Rivers Suyal and Koshi.

Fig. 1.2. View of Almora, with soldiers of 3rd Queen Alexandra's Own Gurkha Rifles, 1895 (*Source:* Wikipedia; the Free Encyclopedia)

Almora is picturesquely situated on a ridge at the southern edge of the Kumaon Hills harbouring a string of summer resorts (Fig. 1.3). Ranikhet, the nearest town, was a summer station for British troops about 40 Km west of Almora. Pithoragarh is located in the Shor Valley in the lower Himalaya region, at an altitude of 5,000 feet. It lies in the eastern part of Almora district, 85 Km east of Almora, at the intersection of well-traveled trade routes from Nepal to Naini Tal and from Tibet to the plains. Pithoragarh is 23 Km from the border with Nepal, and two weeks' travel from Tibet, which lies to the North. The lovely Shor valley is about 16 Km in length and several kilometers wide, and is surrounded by high mountains (Badley, 1931). To the north, there is a fine view of the eternal snows, with Panchuli over 25,000 ft. in the center.

Fig. 1.3. Almora in mid-19th century (*Source*: John Edward Sache (1824-1882)

At present the quaint valley village of Almora, famous for its glorious sunrises and sunsets which offer magnificent views of the valley, is known for its breathtaking beauty and alluring panoramic views of the surrounding Himalayas (Fig. 1.4). Nestled within higher peaks of the Himalaya, Almora enjoys a year-round mild temperate climate. The Chand Rajas established Almora and the British strategically developed it into a fine hill resort since. Till the Gurkha onslaught, in the last decade of the 19th century, Almora remained well protected naturally.

Fig. 1.4. Almora township as it looks now. (*Source:* Wikipedia; the Free Encyclopedia)

At present, this is also famous for being the spot where Swami Vivekananda, the great Indian scholar of Hinduism, had stopped for a break during his quest of knowledge in the upper Himalayas. The journey of Vedas to the west started by Swami Vivekananda in late nineteenth century is still on its way to create new horizons of understanding spirituality. Swami, after a successful journey to the west continued his effort to acquaint the citizens of India. In that era, the people of India suffered sorely under the British rule and barbarism of catholic missionaries to adopt Christianity. His speech at parliament of religions is still imprinted in streets of Chicago (USA) called as "Swami Vivekananda Street". Known as the best known orator monk from India, Swami found his way to develop the milestone mission of Ramakrishna Paramhansa in the Himalayas.

Dr Ronald Ross had a highly distinguished legacy of parentage in armed forces, including that in the British Indian Army.

His father C.C.G. Ross, KCB (1824-1892), a Scottish officer in the British Indian Army, was married to Matilda Charlotte Elderton (1832-1906), a daughter of Edward Merrick Elderton and Charlotte Mayzod Marcella Elderton, in 1856. Both parents were born in India; while mother was born in Jabalpur, his father was born in Saugor (Sagar), Madhya Pradesh, to the illustrious parents Lt. Col. Hugh Ross and Eliza Watson. He joined the British Indian Army in 1841 as a Leautinent and rose to the rank of Major-General in 1881 and full General in 1890. He also fought in the Jowaki Expedition in 1877 when the British government in India proposed reducing the allowance to the Jowaki Afridi tribe for guarding the Kohat Pass. In response the tribe cut the telegraph wire and raided into British territory, leading to two British sorties into its territory. The first sortie of three columns in 1877 was led by Colonel Frederic David Mocatta of the 3rd Sikh Infantry and the second of two columns in 1877-1878 by two brigadier generals, namely, Charles Patton Keyes and Campbell Claye Grant Ross. For his valiant services to the British Empire, General Sir C.C.G. Ross was made a *Companion of the Order of the Bath* in the 1877 Birthday Honours, at which time he was in the Bengal Staff Corps. He later rose to Knight Commander of that Order. He died on 20th June 1892 in Eastbourne and his wife a few years later in 1906, both, in England (Fig.1.5).

Fig. 1.5. General Sir C.C.G. Ross, KCB (1824-1892), Bengal Staff Corps.

Unlike Ronald Ross, his younger brother Charles Ross (10th March 1864 – 21st December 1930), CB, DSO, followed *ditto* the footsteps of their father and rose up to the rank of the Major-General in British Army. He was active during the Boer War and the First World War, where he commanded 6th Division from 1915 to 1917. Charles Ross was the third son of General Sir C. C. G. Ross. He was educated at Stubbington, and joined the Norfolk Regiment in 1884. He was attached to the Egyptian Army in 1893 and 1894, and attended the Staff College, Camberley from 1897 to 1899. Shortly after leaving, he was posted to South Africa, following the outbreak of hostilities there, to act as a divisional signaling officer. He was then assigned to intelligence work on the staff, and remained in the country until July 1902, being mentioned in dispatches and awarded the Distinguished Service Order. In 1904 he was posted to the Royal Military Academy, Woolwich, as an instructor, and in 1905 was transferred to the Royal Military College, Sandhurst as the commander of a cadet company.

He remained at Sandhurst until January 1908, when he was posted to the Staff College as an instructor; he was well regarded as a lecturer by his students. During the early stages of the war, Ross commanded a brigade. On 14th November 1915, he was appointed to command the 6th Division in the place of Walter Congreve, who had been promoted to command a corps. He commanded the division during the Battle of the Somme, where it was engaged in September and October 1916. He held command until 18th August 1917, when he was relieved. Whilst commanding the division, he was made a Companion of the Bath. He subsequently commanded the 69th (2nd East Anglian) Division in the UK.

Like his eldest brother, Ronald Ross, who had an interest in writing poetry, Charles Ross too wrote three books stemming from his academic work: (i) Representative Government and War (1904), (ii) The Problem of National Defense (1907), and (iii) An Outline of the Russo-Japanese War 1904–1905 (1912). He also wrote fiction, publishing at five mystery novels: The Fly-By-Nights; The Haunted Seventh; Every Man's Hand; When the Devil Was Sick; and The Castle Fenham Case. He married Clara Marion Horton, the widow of an officer in the Royal Artillery, in 1905; they had no children.

Thus, the spirit of valour and 'never ever give up' was inherent to Ronald who, during his much troubled career in the British Indian Army, exhibited an extraordinary character of a fighter, rising up through thick and thin time and again to accomplish the task that so far evaded the world researchers – the inextricable link between malaria and mosquito.

"Who would deduce the dragonfly from the larva,
The iris from the bud,
The lawyer from the infant?
...We are all shape-shifters and magical reinventors.
Life is really a plural noun,
A caravan of selves."
— (Diane Ackerman)

• • •

PART II: POLYMATH

2. Schooling and medical education – *building future* (1865-1881)

2

Schooling and Medical Education – *Building Future* (1865-1881)

"होनहार बिरवान के होत चिकने पात"

Transliteration:

"Honhar birwan ke hoat chikne paat"

Translation:

"The geniusness of a person is reflected in his childhood itself."

Ronald was the eldest of ten children of General C.C.G. Ross and Matilda Charlotte Elderton, namely, Ronald Ross, Marion Adelaide Ross, Alexander Ross, Claye Ross, Charles Ross, John Ross, May Ross, Isabel Annie Ross, Hugh Campbell Ross and Edward Halford Ross. Being the first child Ronald was naturally the apple of his parents' eyes. Generally his parents had taken him on long outdoor walks among the woods which abounded in chirping birds and elegant butterflies fluttering their wings galore. Occasionally mounting a four-wheeled cart which was pushed manually, little Ronald noticed some dragonflies appearing suddenly from the nearby bushes along a stream and dancing with an intention to perch on his cart's handle or the hood. He felt very exhilarated with their flight maneuvers and even threw his hands on them to capture, albeit in vain. His father was a spirited sportsperson and had a reputation in running for miles during daily drills. He played many sports and games too, though long running was his forte. Since he loved history books which he devoured in his leisure, he also liked to tell the young Ronald fables of valour from the wars the Great Britain had fought against the French troops of Napoleon Bonaparte, in particular. He even took interest in poetry and art, sketching the beautiful surroundings mystified by dense pine and fir trees, and the wild goat and the women, with babies tied up on their back, at work up the hills. On one occasion Major CCG Ross took wife and young Ronald, still lapped up in the father or mother's arms, to Dhar Chula outpost at the bank of river Kali (which in plains, about 300 km downstream, assumes the name of Saryu). The scene of mighty Kali, flowing with swirling waters in towering leaps with spontaneous high-pitched echoed and re-echoed resonation between the hillocks on the river's two sides, was frighteningly mesmerizing. He quickly took to his pencil and the art-book, and prepared a sketch. For the night they put up a tent (Chowldari) and spent night in the nature. In the full moon light the river looked a perfect cascade of silvery mercury.

On return Major C.C.G. Ross found themselves in the midst of spectacular agriculture fields grown along the slopes of the mountain in a ladder-like manner at a place called Askot, known for its musk deer and verdant valleys carpeted with *Rhododendron* shrubby fields. In the twilight of clear July night the bare mountain slopes shone like silver canopies. The next stop before arriving Almora was, of course, a hill resort by the name of Kausani (to which Mahatma Gandhi later referred to as the "Switzerland of India" due to similarity in landscapes). Kausani is a hill station and village situated in Bageshwar district in the state of Uttara Khand, India and is famous for its scenic splendour and spectacular 300 km-wide panoramic view of Himalayan peaks like Trisul, Nanda Devi and Panchchuli. On the early morning hours Major Ross, staying in a army rest house there, positioned himself with his pencil and sketch-book with the spectacular view of snow-clad Trisul mountain in front of him. Just when he made the first stroke of pencil he heard his wife coming by his side, holding in arms a fully awake young Ronald. Though still very young to understand the nature in the wild, Ronald was smiling all the way and murmuring something. Before leaving the spot, Major Ross had a better look of the majestic surroundings and, while being completely speechless and at his wit's end, pensively thought that if Jehangir, the Mughal King of 16th century India, had visited the place earlier than the Kashmir then he might have certainly extolled the beautiful Kausani as: "*Gar firdaus bar-rue zamin ast, hami asto, hamin asto, hamin ast*", meaning "If there is a heaven on earth, it's here, it's here." Major Ross took his family, including the little Ronald to other as majestic mountainous resorts like Ranikhet and Pithoragarh etc., inculcating into young Ronald a trait of peregrination and its varied benefits.

Largely the environment in Almora remained peaceful albeit ever escalating freedom movement inferno in most parts of the country, especially in view of some agents of the American Methodist Episcopal Society having escaped from the mutiny in Bareilly came to Naini Tal and commenced a mission there for the Society in 1857 (Badley, 1931). By the time Ronald was seven years, C.C.G. Ross, had been promoted to the rank of Colonel and well-settled with their three children. Even though there had been a Christian missionary school, thanks to the painstaking efforts of reverend John Henry Budden (1813-1890) who spent most of his adult life working as a missionary in Almora, where one of the oldest churches was renamed in dedication after his death in 1890. The Budden family worked in the Almora/Pithoragarh area for over 80 years, first under the direction of the London Missionary Society, and later as part of the American Methodist Missions. They founded several schools, medical dispensaries, and missionary hill stations (including ones at Pithoragarh and Champawat). To this missionary convent school mostly children of army officers, but also of a few rich civilians or local landlords, the *Zamindars*, were enrolled for early schooling.

In 1850, the Reverend John Henry Budden was stationed with the London Missionary Society at Mirzapur, having come to India from London in 1841. He had visited the Himalaya hills for his health, and there met Captain (later Sir Henry) Ramsay, who was a Christian officer then resident at Naini Tal. The same year, Revd. Budden was offered a missionary position in the Kumaon Hills by Captain Ramsay and Mr J. H. Batten where the formal opening of the mission work took place on August 7, 1851. It was also decided that a school and other mission work should be begun as soon as possible in Almora. Two buildings adjoining the Bazar (formerly a mess-room and a billiard room for the regimental officers) were dedicated to the mission work. One of the buildings was used as a Mission Chapel, where Sunday services were offered in both Hindustani and English, while the other building was used for a school. The school building was later determined to be inadequate, and another building was secured in 1851 with Mr Rebsch as the schoolmaster and mission assistant. In 1852, an Indian headmaster was appointed for the Almora mission school. With such an arrangement Colonel C.C.G. Ross was little hesitant to send Ronald Ross to a school where the headmaster was a local Indian fearing inadequacy of standard of teaching – a kind of an education phobia then prevalent in most of the British assigned duties in India. In 1871, the mission school was relocated to a more spacious building in the center of town, on a site formerly occupied by the palace of the Almora *rajas*, and later by Government buildings (Badley, 1931). In 1886, the high school became known as Ramsay College and was affiliated with Calcutta University (Lovett, 1899; Badley, 1931).

The mission also supervised a leper asylum in Almora, one of the oldest institutions of its kind in India devoted to the service of the leprosy patients under initial care of John Henry Budden, which had been established by Sir Henry Ramsay, then a Civil Officer, in 1840. This sacrificial service took an ugly turn when the management of the missionary began to use their influence to bring about conversion of the impoverished patients, more or less all of them being Indians, to Christianity. The conversion of native Hindustani to Christianity was an invisible objective of the British Government to gradually make their presence felt in India which was constantly opposed by the locals as the country was struggling hard to break open free from the shackles of slavery for over two hundred years by the British! In 1864, the first conversion of a leper was recorded, with over 700 lepers converted to Christianity in the Almora mission asylum between 1864 and 1927.

The early schooling of Ronald Ross was essentially done at home, mostly under the watchful and loving solicitude of mother, but father too spent as much time as possible with the children, as in vogue in those days most British officers had plenty of leisure time following fixed office hours during the part of the day. He was often seen sketching something or the other, even of himself by looking at the mirror. Looking at his gifted prowess in drawing figures his father was so influenced that he would occasionally invite him prepare an extempore sketch of the guest at lunch or supper. Ronald also loved to chant the rhymes mother had taught him about. All along, though Ronald was seemingly pretty good in learning basics, his father C.C.G. Ross was getting restless and worried thinking about lack of formal education for Ronald. He considered good higher education was imperative to develop all round wisdom and eventually decided to give education to him back in England which at that point of time was lauded as a country where the Sun did not set, implying the Union Jack flying high in almost all parts of the world! He had many contacts in England and one such was Mr Captain Barwell who accepted to escort young Ronald to live with his uncle and aunt in England. When Ronald learnt of this development he was heartbroken since he was already in love with Almora's infinite scenic, calm and magnificent environs. He felt born-free there. Somehow mother realized Ronald's indifference and spoke with her husband on the issue. Senior Ross was however adamant and resolved to send Ronald to England for better education prospects. He was actually dreaming to see Ronald one day wear the British Indian Army officer's uniform just like him and continue his family's legacy.

Thus at the age of eight Ronald Ross was sent to England to live with his aunt and uncle on the Isle of Wight where he was enrolled in the Primary School at Ryde. Ronald proved to be an average student, although the class teacher always appreciated his natural instinct to sketch a drawing and play music. He happily participated in school dramatic and cultural activities, with occasional sparks of brilliance in certain disciplines. He, however, kept mostly away from the field sports and games. After passing schools he was sent to a boarding school at Springhill, near Southampton for secondary education, in 1869. Here too he manifested his early childhood passion for poetry, music, literature and mathematics. In mathematics particularly he, at the age of fourteen years, won a book titled *Orbs of Heaven* in prize in the annual school quiz competition which further stimulated and sparked his interest in the subject. In 1873, at sixteen, he secured first position in the Oxford and Cambridge local examination in drawing. His father was regularly following Ronald's progress report in the school and thought it was the right time to let young Ronald know his own personal wish to become an army officer in India. In those days the British Empire was spread in many parts of the world and his father, now a Major General, thought India the best place for his son where great prospects of promotion and hassle-free life prevailed!

General Ross summoned his son and spoke what he thought about his future. In a way he had declared an order because he had brought up his children in a way as hardly anyone ever could muster in enough courage to oppose his decisions. In fact nobody could raise head in front of him and look into eyes straight. Ronald wanted to become a writer and, therefore, much hesitantly put his thought before his father. Hearing his son's wish to become a writer his father was sorely infuriated and shouted at him scolding for having wasted all his valuable time and money in learning a lowly placed vocation. He roared and pronounced his verdict that Ronald would go to get enrolled in British Indian Army. Ronald was disappointed but was in no position to oppose his father. His mother, who was so far witnessing the entire episode rather meekly, stepped forward and intervened by coolly supplicating the young Ronald and making him comprehend his father's concern for his own future betterment. Ronald argued with her but failed to see any change in her stand. Reluctantly though, he finally consented to his father's wish.

General Ross initially ventured to send him to get enrolled to the British Indian Army but finding neither Ronald's temperament nor the aptitude up to mark for the coveted job, he, using his lofty stature and wide contacts, secondarily arranged his enrollment at St. Bartholomew's Hospital Medical College in London, in 1874, so that he could join the Indian Medical Service (I.M.S.) under British Army in India. Before Ronald packed his bag and baggage to join the medical studies, his parents deftly sermonized to him to work hard and prove his worth lest he would be left to repent all his life! Ronald was, however, too, young to comprehend all that his parents talked about India, but somewhere in the back of his mind he certainly felt happy that he would be at least away to a new place and country called India, and use his time the way he wants. There at the Hospital, too, Ronald did not have many friends and mostly kept to himself with flights of poetic and artistic imaginations. He always carried with him a pen and a copy to jolt down his thoughts. as they sprang in his mind. He wrote about the human nature and made penciled drawings of anything that struck his imagination. Consequently, he was never really fully committed to his medical studies, and consequently cleared the examination mediocre performance, in 1880. But this was not enough to join the Indian Medical Service.

Later, as a necessity, he passed the Society of Apothecaries examination at the second attempt, and his performance was such that he received in 1881, only after an additional four-month training at Army Medical School, a commission for the Madras service, considered by many the least prestigious of the three Indian Presidencies; Bengal and Bombay were the far more desirable appointments by the incumbents. With a heavy reluctance he accepted the posting keeping in mind his father's stern advise. He had overheard way back in England about the fables of all the world' leisure at the disposal of Britishers serving in India and enjoying by whiling away their times mostly fishing, golfing, hunting and amassing wealth. Anyway, he himself nurtured the thinking that this forced 'exile' to India would end before long and he would return sooner to England for a much comfortable future than in India which he then knew as 'a country full of dust, heat and mosquitoes', never ever realizing in the remotest sense that the same country would one day raise him "from nothing to an all-important man" and reign the world of malariological science and be honoured with the world's highest recognition, the Nobel Prize in Physiology or Medicine (1902)!!

Fig. 2.1. Dr Ronald Ross. (*Source*: SRRIP, Archive).

"Life is a matter of choices,
and every choice you make makes you."
— (John C. Maxwell)

● ● ●

PART III: MOSQUITO-MALARIA CONNECTION

3

India, The Land of Great Virtues and Opportunities: Ross's "Karmabhoomi"–*First Steps into the Habit of 'Self-Culture'* (1881-1888)

''जहाँ डाल-डाल पर सोने की चिड़ियां करती है बसेरा
वो भारत देश है मेरा''

— (Rajinder Krishan)

India, the 'Golden bird' of yore, with its deep-rooted traditions, lofty vital qualities and diverse languages influenced by a history that is several millennia old and spanning across the Indian Subcontinent, is one of the oldest civilizations in the world with a kaleidoscopic variety and rich cultural heritage which magnetically attracted explorers and entrepreneurs from all over the world, especially the Great Britain and Europe. A country with a distinct and yet diverse geographical entity, it is bounded by the Great Himalayas in the north, from where it stretches southwards and at the Tropic of Cancer, and tapers off into the Indian Ocean between the Bay of Bengal on the east and the Arabian Sea on the west. However, following the continuous barbarous quests of Moghul invaders in the 10th century onward and later by the British, among several other Europeans – who all were permitted to land and begin some kind of petty trades of mutual benefits, eventually treacherously grabbed not only most of the market first but also wielded their tyranny and reigned supremacy over the land. During both these periods of foreign rules, *i.e.*, Moghul and the British, India and her people were sorely impoverished in health and economy, and their cultural heritage abominably damaged systematically, but in vain! During the mid-19th century, it was generally eloquently advertised that the Sun never set on the British empire, while the English tradesmen under East India Company were busy in scooping with both hands the wealth of India and considered the country a safe haven for their hideous business. Sooner the British Empire, in particular, taking stock of wealthiness of resources in India through its East India Co. tradesmen, began to visualize a rather glittering and grand future of her own crown in United Kingdom. The British government in due course of escalating British influence in India drew an extensive plan to reinforce extensive military services in the whole of Indian mainland with an underlying objective to expand and strengthen their rule for a far distant period. The roadmap to plunder Indian richness was ready but the road ahead was not easy for the whites since, in addition to a string of heavily intensified towering protests engendered by the Indian populace for freedom from the greedy hands of the British, malaria and mosquitoes, too, stood their way to progress!

Despite all the aforesaid prodigious splendour, India, however, was also a heartland for malaria that for centuries not only emaciated prospects of harvesting the natural resources and impoverished the intelligentsia, it also sorely affected health economy of her people, apart from exacting a heavy toll of lives and affecting millions of people annually. Unlike in South American country of Peru where quinine, the wonder antimalarial drug was crudely consumed after extraction from *Cinchona* bark, there was no such panacea available in India albeit some time-tested domestic treatments mostly including plant extracts such as Shyama Tulsi, *Ocimum* sp. Following the epoch-making discoveries about tuberculosis bacillus by Robert Koch in Germany, most European countries, but particularly Great Britain, eyed the escalating malarial inferno in India as an opportunity for its fast emerging pharmacological industry and biomedical science, besides a driver for carrying expansion in the region. To the British Empire, India was the true land of out opportunities and growth!!

The intersection of Britons with India in the role of colonists was necessarily attended by considerations of health. From the seventeenth century itself, hospitals had been established in the colonies within close range of trading and military posts. The Bengal Medical Service was created in 1763 with an initial complement of 40 surgeons. This model was replicated in Madras and Bombay, leading in 1785 to a decision by the Court of Directors of the East India Company to set the number of surgeons and assistant surgeons for the three presidencies at 234, a number that would rise to 630 by 1824. The eventual result was the formation of the Indian Medical Service (IMS), which met the needs of both civilians and armed services, such that doctors associated with the Service were also military men. It was axiomatic that the IMS would be staffed in principle by Britons, whether sent from the United Kingdom, or as products of British parents based in India. Dr Ronald Ross (1857-1932), whose malariological explorations became the basis of the 1902 Nobel Prize, was an example of the latter.

The reason why malaria became the focus of research was not the same for all of the individual scientists and countries engaged in the quest for deciphering the enigma behind the disease, that is the causal agent of malaria – colloquially known to have associated with the 'bad air' from marshes, among others! Two countries, France and the Great Britain had stood out in encasing foremost the prevailing situation and expanded their colonies throughout regions of the world where malaria was the most serious and debilitating of the many parasitic tropical diseases, a factor that demanded limited expenditure since in most of these places or countries and regions French and British Empire already had their dominions and could freely deploy their scientists and/or servicemen also as a tool to reduce burden of job opportunity. Americans made a rule to send their people to India on a permanent basis for life long. Therefore, prominent scientists decided to tackle the problem of paludisme in France, Germany and in Great Britain who were either posted in the malariated countries or regions or had some links there. Italy, as an European nation, had a unique situation; malariological research thrived there, both from the clinical and the public health points of view, but also because Italy encountered a serious malaria situation in the southern region of the country where disease was endemic and Rome, the capital city, was prey to malaria in the summer and autumn months which was also the period of maximum reproductive activity of the mosquitoes. Obviously malarial studies had a different perspective for all these countries: for French and British researchers, malaria was a "tropical disease" since both the countries were free from infection on mainland, and they mostly thought it a "colonial" problem, whereas for Italian physicians, malaria was a persistent problem to their country and a scourge that hindered the development of many southern areas which recently achieved strategic national unity of the country.

The reality behind any humongous achievement in life is that it is accomplished through dint of hard work with focused attention for many years, or may be, whole life! Accident does very little towards the production of any great result in life. Though sometimes what is called "a happy hit" may be made by a bold venture, the common highway of steady industry and application is the only safe road to travel. It is understood of Dr Walter Reed that, as trained in the army way of exercising discipline in every walk of life, he meticulously cultivated the habit of examining his material of study again and again until he had comprehended scrupulously his matter. This is how he arrived at the final conclusion of his several year-long investigations on the Yellow Fever in Central and South Americas and narrowed down decisively on *Aedes aegypti* as the vector for the deadly disease. The capability of putting in these all wisdom is acquired only by the labour of a life; and the probability is, that a scientist who speaks about his results without verifying critically by repeating the experiment several times would only put himself to remorseful ridicule and repudiation at the hands of others, at last. Sedulous attention and painstaking industry always mark the true worker. The greatest men are not those who 'despise the day of small things", but those who improve them the most carefully. They neglect nothing. Great Manson's exhaustive research on human lymphatic filariasis in China is an extraordinary example of ceaseless industry and observational subtilities. He left no opportunity to find the filaria worms in live or dead human bodies as well as mosquitoes, albeit personally devoid of any basic knowledge about the insects.

Dr Ronald Ross's assignment in India as a Surgeon in the I.M.S. on 2nd April, 1881 began with a training in the Royal Victoria Hospital and School at Netley for the prescribed course of military medicine and surgery, lasting four months, before sailing to India later the same year, and was actually nothing very special in the Nature's eyes until, of course, the destiny began to exercise future course of events in his much tumultuous life. He worked hard for the examinations at the Hospital hoping to better his place in the concluding examination so as to have a choice of presidencies in India such as that of Bengal, but he actually went down two places. Much against his own desire he however he got posted to the Madras presidency, the least liked by Ross. He boarded the troopship *Jumna* on 22nd September, 1881 and, after a month of sailing *via* Suez Canal, arrived at Bombay (now Mumbai) on 23rd October, 1881.

In Bombay an arcane incident occurred the very next morning in the hotel Ross was staying in; he was woken up to a quaint avian welcome by a crow's punctuated cawing from inside his room's window. He did not find anything particular in the bird's yelling, but customarily it alluded toward a guest's prospect of arriving coming back to! This was rather strange since no other foreigner, particularly a Britisher, had ever been confronted with such a scenario. Thus, Ross was given a warm-blooded (because avians are homeothermic) welcome to India, the country which would later take him in his prime time of career to the pinnacle of fame all over the world, once again, through his research on such avian life only! But, young Ross was not superstitious and he found nothing an unusual for Indian crows (*Corvus splendens*).

Between 1881 and 1894, with the base camp at Madras, Dr Ross was *inter alia* ordered to temporarily serve in Burma, Quetta, Baluchistan, Andaman islands, Bangalore etc. to perform various organizational responsibilities, in addition to medical issues, administrative and other matters of military nature.

A day after landing at Bombay, he boarded a train for Madras, reaching there after three days of a very tiresome journey. He however had very pleasant memories of the lush green prairies and agriculture fields, heavily bouldered landscapes, rivers, forests and local folks with simple clothes. Soon after arriving his regiment he was escorted by his orderly, named Appoo, to his accommodation in Dent's Garden Hotel, located on a stone's throw from his office. Appoo, a native from south, was as hard-working, faithful, docile, and intelligent as any other Indian. He remained with him until his last sometime in 1894 when Ross away to his home in England on the first furlough. Ross remembered him as a very disciplined and caring human being, and respected his workmanship. Though the Hotel was a large old house, in fact, with great lofty rooms, plastered walls, large verandas, cool bathrooms, with which Ross was indeed very pleased, but he also found himself surrounded, almost whole day and night, by the mosquitoes - his lifelong subject. This was his first encounter with the mosquito.

At the Hotel he had a mosquito-net drawn along a squarish wooden framework to cover his bed during sleep in the night but the sound of mosquito flight and fear of their ambush through net-holes kept tickling into his head all through. He never imagined that destiny had designed for him to research on these very critters all through his life and the same tiny animals would one day make him world famous; [As a convention, Indian scientific community celebrates Dr Ross's award-winning mosquito research as 'Mosquito Day' on 20th August every year in his memory; and they celebrated the Centennial Year of his discovery depicting mosquito-malaria connection on the 20th August, 1997 by organizing an international conference in Hyderabad (*Organizer*: Dr V.P. Sharma) and, of course, they also celebrated his 150th Birth Day, as a 'Tribute to a great scientist and Nobel Laureate' during a one-day seminar about 'Ross Perspective: Global Warming – Malaria Alarming', on 13th May, 2008 in the historic building in Secunderabad, where Dr Ronald Ross had discovered "malaria parasite's oocysts on the gut wall of a 'dappled-wing' mosquito' now preserved as the Ronald Ross Institute of Parasitology under the care of Osmania University (*Organizer*: Dr B. Redya Naik; Chief Guest: Dr B.K. Tyagi, Fig. 3.1a, b).

Fig. 3.1a-b: (a) celebrated Sir Ronald Ross's 150th Birth Day (those seen include, from left, Dr B.K. Tyagi, Prof. Surender Rao, Dr P. Jambulingam, Dr K. Raghuvendra, and Dr B. Reddya Naik), (b) Garlanding the bust of Dr Ross as a 'Tribute to a great scientist and Nobel Laureate' during a one-day seminar about 'Ross Perspective: Global Warming – Malaria Alarming', on 13th May, 2008 (those seen include, from left, Dr B.K. Tyagi, Dr Kareem, Dr. A. K. Chari). (*Source:* Dr B.K. Tyagi Archive).

At Madras Ross was posted to the Station Hospital (Fig. 3.2) meant for British troops only as there was then no vacancy in the Indian regiments for the newly arrived doctors, but also the promotion was blocked for them, as Ross himself said, "for some years by incompetence of someone who had overcrowded the junior ranks.' Here he was a junior and temporary officer (he was made permanent in 1990 only).

Fig. 3.2. Major Surgeon Dr Ronald Ross (*Source:* SRRIP, Secunderabad, archive).

As for most of the young British officers, with only a couple of hours duty every day, Dr. Ross too had plenty of leisure time, golfing, fishing and hunting, and only dutifully, but pleasantly, performing his medical duties generally for 2-3 hours in the forenoon but occasionally a round to the wards of patients in the evening. He maintained a good rapport with his seniors at the Hospital. He did not purposely resort to doing regular exercises like his many other unit mates since, he felt, he was already walking briskly a lot up and down the stairs of his hotel as well as office corridors. Nonetheless he enjoyed long walks in the early hours' cool breeze drawing from the majestic seashore line nearby. A voracious reader Ross used to go all around the Madras City in quest of books but his favourite was M/S Higginbotham's book-shop where he got all his literature by, such as for example, Tasso, Alfieri, Racine, Hugo, Schiller, and Goethe, besides Homer, Aeschylus and Virgil. He was an early riser so he would completed as much reading and composing scripts of poetry, novel etc. as possible for him in the early morning hours, before the military bands at early parade floated in the air, and slowly but certainly the sun – his 'watch tower in the east' - came also. Living in solitude was difficult but proved a boon for Ross for this was a time of integration of thoughts and experiences, spent with a wonderful environment imbibed with literature, reading, writing, composing theatrical essays, painting, swimming, and, more than anything, good old friends.

In those times passing a lower standard examination in Hindustani was obligatory for all incumbents for military service in India. Ross had an advantage because he was born in India and much of his childhood, till he was almost eight years of age, was spent here. His father had many servants, all of which of Indian origin, from the civil side of the military force who were instructed to carry out many different household jobs, including also to care the child, Ronald Ross, and his younger siblings. These house servants, called orderlies, spoke Hindustani among themselves, and also with the little Ross. They rhymed the fables of valour about Lord Ram, Lord Krishn, Mahoba's valiant Aalha-Udal brothers and sacrificial love story of King Nal and Damayanti. Ronald's father appreciated and approved of all these rhetoric teachings to the young one for he himself was a deeply learned person and very well knew the positive effect of these great epics on the growing mind of the child. The little Ross too became friendly with them and pleasantly responded his babysitter in Hindustani. In fact, Ross was proving very good in communicating in Hindustani with all including parents and siblings in and outside home. Till about the age of five years young Ronald Ross carried very good health and by then had learnt reasonably well about spoken Hindustani from the attendants. But once he fell sick due to typhoid, he became not only severely emaciated in health but also got disconnected from the company of Hindustani attendants. The depleting health and impending necessity to provide him with good education, in the tradition of all the Ross Family members, his parents soon arranged to send him to England. As would be expected Ross in course of time forgot completely what he learnt in Hindustani back home in India from the native 'tutors'. However, seventeen years later when Ross was destined to clear an examination in Hindustani, to raise his future career prospects in the I.M.S., he felt initially that he had utterly forgotten the language but as he practiced more it began to come to him rather naturally. He was amazed about the blitzkrieg pace with which he took control of the spoken Hindustani, just as German, Italian, French and some other European languages happened to come easily to him. He passed the required examination in Hindustani on 9th January, 1882, with a period of less than four months of his reorientation on the Indian soil, thanks to his early companionship to those phonetic Hindustanis.

Later, he moved to official residence which too had plenty of space to stroll. Being a bachelor yet he did not mix up with many there, although plenty of requests for dinner served on him. In Vizianagram (or Vizianagaram app. 21 Km from Vizagapatam or Visakhapatnam, in Andhra Pradesh, on the east coast of the Indian peninsula about 300 Km north of Madras; in those times to be reached in 2-3 days by steamer), where he was moved from Madras on his return from Bangalore in March, 1882, he got pretty well acquainted with the all pleasant Indian civilian life, the few officers of the regiment and their ladies, the young, sportive and entertaining Maharaja. All were very kind to him and engaged with him in varied sports which he feared he could never become a good shot owing to his slow co-ordination of eye and hand. Always, without a lapse though, Dr Ross was extremely grateful to all such offers but because he needed time to complete his self-awarded projects, in different subjects, of course, such as poetry, novel, arts or drama, mathematics etc., he would politely express his inability on one or another pretext. All his colleagues and friends knew well that, though he easily socialized during the official parties, dining, dancing and drinking wine to give company to fellow officers he was not a heavy drinker at all, but he smoked cigarettes – a habit he could never give up. On the other hand, Ross was good at certain arts of the sportsman usually going out shooting, playing tennis and billiards. Apart from sacrificing such times in the name of 'compulsory' social gatherings, as in vogue among the ranks in military, he was left to himself for most part of the day. He took up many ventures simultaneously including a long-standing interest in mathematics. He finished his first novel, as well as writing some poems and short plays but could not complete many other literary pieces which frustrated him much. He began to realize that there was something else more important he must embark on. His own past-thinking that he would have all the world's leisure in his posting under Indian Medical Service was slowly getting eroded and he actually yearned to act and achieve something more towering in life!

"What ?", he failed to figure out clearly and felt terribly confused; his journey was becoming directionless like a ship in the sea without an idea of its destination! He had unflinching confidence, but he was longing to apply his energies in the 'right direction', to fruition. He felt he might lose the bus and could never achieve for which he was longing all these years - the glory!

""Real glory, Springs from the silent conquest of ourselves,
And without that the conqueror is noughtBut the first slave."

During his central posting at Madras he visited many other places on duty. About February/March, 1882 Dr Ross was made Medical in-Charge, 10th Madras Infantry, stationed in a camp near the Mysore railway terminus, for a few weeks. The climate here was so pleasant and air naturally filled with fragrance of sprouting seeds and blooming flowers from the medicinal plants aplenty in surrounding forests in high mountains. The vigour in Ross was so high that every morning he woke up early and walked long distances among woods and rocky crests of mountains, fir plantation, dells with beautiful little trickles of streamlets, just to be in time to see the "morning stars singing together." He was completely mesmerized with the wonder of nature and thought only of these things; he seemed to have utterly forgotten every other thing save for the elegant scenery permanently planted in his memory which he readily noted in his diary, his inexorable companion. Ross was beginning to understand the country and the people through his umpteen morning rambles. On this impression of his he wrote in "Indian Shepherds".

In August, 1884, Dr Ross wired to join as Medical in-Charge of the pensioners' depot at Pallaveram (or Pallavaram) in south of Madras where a certain number of British officers and warrant officers who, many with families, had chosen to stay back in India after retirement. The pleasant Indian locales and the society of Pallaveram did him much good. Ross could stay in Pallveram only for a few weeks since we was asked to move out to take over as the Acting Garrison-Surgeoncy at Bangalore, his second assignment since 1882, for six months commencing 6th September, 1883

Bangalore, like Madras, was a large city at the center of the peninsular India, but unlike Madras which is hot and humid, it's climate, being located at about 2700 feet above sea level, was comparable to that of Italy; fairly hot and dry during summer, much cooler with early rains in May, and delightfully cool and clear in winter. The plateau appeared verdant all over in Bangalore, hosting the majestic Cubbon Park, the elegant Lal Bagh Garden, Club, endless games, and a pleasant military social life . It fell within the territory of the Maharaja of Mysore and enjoyed both civil and military settlements. In the south-east lay a very considerable native town of about 90,000 people, close to huge Ulsoor Tank. The entire town was situated on the gently undulating slopes of the flat lands, ensuring quick drainage of monsoon waters. Ross painted a large, clear and admirable picture in his memory but, unfortunately, for some mysterious reason, the salubrious climate did not suit Ross. Yet Ross preserved a picture of pleasure on his mind; the sun, the gardens and pillared villas, the exquisitely dressed natives, and the open-air, aristocratic life of the British – as goodly a place, probably, as there was any in the world. Ross's duties included looking after as Medical in-Charge of the General and his staff, of a number of pensioners, and of several hospitals and dispensaries. He regularly made rounds to the wards to see his patients who explained their certain complaints particularly about the mosquito nuisance in night. Dr Ross was lodging in a sprawling house close to the parade ground or the '*maidaan*' near the Officers' Mess at the 4th Madras Pioneers and it was here that he felt 'interested' in the mosquitoes. He discovered that many mosquitoes were breeding in a small tub and similar other objects just outside his residence and the mess building. He overturned and/or vacated all these where after he did not have mosquitoes troubling him at all. Ross was indeed very happy with his 'discovery' and immediately brought this to the notice of his superior who did not only ridicule him but scornfully instructed to not meddle with the nature. "It would be upsetting the order of nature, and, as mosquitoes were created for some purpose, it was our duty to bear with them" he instructively sermonized Dr Ross to cut short his enthusiasm. Now, this was the first time Ross entangled in a fruitless confabulation with a superior officer but he realized before long that 'this type of fool' was very common indeed.

"If they answer not to your call walk alone

If they shut doors and do not hold up the light
when the night is troubled with the storm,
with the thunder flame of pain
ignite your own heart
and let it burn alone."

— (Rabindranath Tagore)

Dr Ross received an order in September, 1884 to proceed to Quetta, in Baluchistan, and tasked with bringing back the injured of the 1st Madras Pioneers. His orderly, Appoo, accompanied him all through, and Ross, on his part, cared for his requirements especially in chilled nights in and around Quetta. On one such night when Appoo shivered in sleep he provided extra warm clothes and blankets. The area around Quetta was richly verdure, full of pasture, and Ross was delighted to feast on meat, especially mutton, 'the best he had ever eaten in the East'. His peregrination to Quetta proved very challenging though enjoyable. Owing to their tall and strong built, Ross felt a great interest toward the Baluchis who, in his opinion, were either very handsome or else raw, savage and murderous!

On his return from Quetta, he served as the Medical in-Charge for some months and was then transferred to Burma (Myanmar) in January, 1885, to take medical charge of the 9th Madras Infantry stationed at Port Blair in the wonderful Andaman islands, situated some 750 Km southwest of Rangoon (Yangon) and 1200 km east of Madras. Ross was happy because he was paid an additional allowance of Rs. 100 a month which he did not draw while posted at Madras. Appoo, his servant, was this time not allowed to accompany him. He took charge in Thyetmyoo and moved to Port Blair by steamer. In those times British used the island of Port Blair to confine in jails on shore many hundreds of transported Indians who had raised voice against the British regime's atrocities in the mainland! Britishers called them 'convicts' who had never any criminal record and could not have even killed a fly or a mosquito!! Many of those 'convicts' who did not kneel before the continued British torturing in jails or convict settlements, were unceremoniously plunged into the turbulent sea in the middle of night, very many times. But, all the British officers, their families and other whites in the army garrison lived comfortably in the luxuriously made wooden houses with elegant gardens on another island, the Ross island, about a kilometer from Port Blair. He stayed there for three months, enjoying largely the beautiful gift the nature has awarded to the archipelago in its ambience, the fauna and flora. Due to strategic decision once again Dr Ross was ordered to arrive Moulmein which he did via Calcutta on 1st May to arrive on 13th May, 1885 and which he was extremely pleased with as he exclaimed, "One of the most beautiful stations in the world." Appoo was back with him in Moulmein. Here Ross lived with his old friend, Captain Kirkpatrick, adjutant at the 9th Madras Infantry and his expensive bungalow was mostly at the disposal of Dr Ross who had all the world's leisure to do what he wanted, i.e., whiling away in varied sports and indoor games, besides his exercises with mathematics, poetry etc., and of course the little medical urgencies. In Moulin he came across many such patients who were in urgent need for surgery and or prolonged medication. He encountered a large number of patients suffering with the severe infection in their intestines by the roundworm, *Ascaris lumbricoides*, which he treated successfully with a few doses of santonine and caster oil. This made Dr Ronald Ross quite a famous name among the patients. Even this pleasure time proved ephemeral as he was once again instructed to leave for Port Blair, to the 9th Regiment of the Madras Infantry, in May, 1886.

In Andamans, after attending to his scanty medical attendance in the morning hours he would, as a rule, go out for long pony-rides through woods or sail in see fishing almost whole day defying unsuspecting emergent weather with heavy downpour the fear of getting drifted away from his direction. He admired beyond comprehension the virgin beauty of the Andamans. Actually he enjoyed being in the Andamans where he had found plenty of time to complete his long-pending literary composition and mathematical exercises.

But these happy times ended suddenly when the 9th Madras Light Infantry was ordered to Madras City early in 1887. Ross had to sell out all his such dear belongings as boat, but completely refrained from bidding his beloved pony, violin and piano. The pressure of constant travelling without a significant sojourn at a place, temporariness of service and loss of additional allowance that he was drawing in Andamans was becoming unbearable to Ross and he felt a shattering effect on his health and workmanship.

Hardly had Dr Ross arrived Madras, he was shown way to another native regiment there, the 3rd Madras Infantry, stationed at Royapetta. Time here was his own, as usual after serving to patients in the regiment hospital. He, along with other fellow officers of the regiment, went to fishing in Cauvery and Bhawani rivers, the latter flowing out of the Nilgiri Hills hosting Coonoor and Ootacamund – both heavily malariated. He made as many as eight visits to the spot, angling throughout the day, braving hot evenings, and returning home by twilight on foot. In between explorations he solved mathematical problems with an objective to bring in his own theory, but in vain. He felt as if his mind had stopped to work any more at anything! He then changed his attention to something never attempted earlier- Chemistry. This too proved of no avail to normalize him. Deep inside his heart he began to feel becoming hollowed, quietly and progressively. In fact, against his wishes and principles, he had resorted to heavy beverage-drinking to overcome his rousing depression. Suicides by a few officers in the regiment who suffered homesickness was beginning to have a toll on his mind, too, and his solitude or bachelorhood he once considered a blessing in disguise was in actuality turning into a bane. Due to this his health and vigour rapidly deteriorated but he could not single out a cause behind it all. He wanted to run away to some distant place where he could be happy! These were like black days of his life, for the first so far. He pondered, "What was his life's motto in India?"

"This above all,—To thine own self be true;
And it must follow, as the night the day,
Then canst not then be false to any man."

Dr Ross had been arduously toiling without any rest or, almost, a day's medical leave, and so a kind of a incomprehensible illness fell on him which seemed to devastate him from both outside and inside. He could no longer work, nor even play instruments or read books, leave aside saddling his dear pony. He was fast losing confidence in himself and his potential, and that his perseverance was going unseen or unrewarded. In Dr Ross's own words," Hc was overcome with the horror of *cui bono*." Of late he it also dawned on him that he was possibly not giving full attention to his medical responsibilities. He had heard of great discoveries of microbial pathogens of cholera and tuberculosis, and it pinched him much that he was still fruitlessly longing in dark to equal his marvels. The doctor inside him for the first time voiced, though in a low nearly dead pitch, to strive hard with a focus toward his noble profession, to discover the causes of those diseases which were perhaps mankind's main enemies. He had a feeling that he was living in a cemetery.

'tis not in mortals to command success,
'll do more, deserve it.

Out of frustration, anxiety and insecurity, Ross, who till now was an epitome of calm and composed personality, showed for the first time signs of exasperation and consternation, and, towards the middle of 1888, applied for grant of furlough due to him, with a rider of course; he threatened the administration to resign if the leave were refused. The one-year long leave on furlough was granted without delay and he immediately departed, via Colombo (in Sri Lanka) for England by a ship on 28th June, 1888. He sketched his on-going state of mind in his *Philosophies* ('The Brothers' and 'Alastor') during the voyage. Ross in his 'Memoires' has commented that 'he could endure India no longer' which, on a careful perspective analysis of events between 1881 and 1888, does not seem to hold water simply in view of the four facts, *viz*., (1) Dr Ross was an officer in the Indian Military Service with all the comforts, not to be seen in his own fatherland (England) particularly in case of a very junior officer of his level, such as for example availability of a 24x7 personal orderly (servant), (2) provision of an official residential quarter with all basic facilities, (3) of the 24 hours of the day he was required to care his patients for not more than a couple of hours only, and the whole remaining time he was free to utilize to varied matters of his interest that regularly provided him a moment of introspection, and (4) He maintained throughout a cordial and congenial relationship with all the officers in the regiment, including their families, as well as the servant, Appoo. But, yes, Dr Ross nurtured a feeling of biasness on the part of the government in paying less salary by Rs. 100 a month all through his posting in Madras; for transferring time and again to sometimes far flung areas, and for not considering his ceaseless toil and moil for any reward or promotion. Therefore, if Dr Ross suddenly felt disinterested in continuing his career with the I.M.S. it was only due to 'fearful' feeling of "home sickness" that in fact many of his fellow officers had suffered before his eyes, and he was surely affected by those events.

"The voyage of discovery is
not in seeking new landscapes
but in having new eyes."

●●●

4

Being Polymath: A Sure Way Forward in Charcter Building – *A Genius is Self-Made*

"Learning gives creativity
Creativity leads to thinking
Thinking provides knowledge
Knowledge makes you great"
— **(Dr A.P.J. Abdul Kalam)**

Before unfolding the marvels of 19th century's greatest discovery, it is considered opportune to have a look on some of the important 'characters' of its discoverer, Dr Ronald Ross, who by sheer dint of hard work and focused attention achieved his objective. This will clear many doubts in the mind of readers who have earlier developed a conflicting opinion about the working attitude of this great man, a true genius nonpareil.

Though a son of an army General Dr Ronald Ross never ever sought any favours either from him or his acquaintances in the British Indian Medical Service. Save for help and guidance of Dr Patrick Manson, his mentor with whom he developed a teacher-taught relation in scientific matters alone, he moved entirely on his own without ever blaming any one or the circumstances. He was a true genius who created his own means and ways to achieve the objective. There is no set definition of the genius of a scientist, but all those who are bracketed within the specific class do have few characteristics common in them such as the continuous thin-king, focused attention, obsession, perseverance and application, and, on the top of all, 'pursue objectives and never think success.' Fortune has often been blamed for her blindness, though the fact is it is man who is blindingly ignorant of the inexorable laws of life. Fortune is generally on the side of the industrious, as the winds and waves are on the side of the best navigators. In the pursuit of even the highest branches of human inquiry the common qualities are found the most useful – such as common sense, attention, application and perseverance. A genius could boast of many things but even they can not disdain the use of these inevitably indispensable, and yet ordinarily looking, qualities. The very greatest men have been among the least believers in the power of genius, and as worldly wise and persevering as successful men of the commonest sort. Some have been defined genius to be only common sense intensified. Genius of man can be defined in different ways as the 'power of making efforts', 'power of lighting one's own fire', and 'power of patience'. The extraordinary results effected by dint of sheer industry and perseverance, have led many distinguished men to doubt whether the gift of genius be so exceptional an endowment as it is usually supposed to be. It is a well-judged thought that it is only a very slight line of separation that divides the man of genius from the man of ordinary mould: "*Extraordinary men do not do extraordinary things, they do things extraordinarily.*"

Once Dr Ross understood the meaning behind message of Dr Manson's dictums, "Follow the Flagellum" and "The Grand Induction Theory", he always thought up to them: "*Dreams are not those which you see in sleep, but those which can not let you sleep.*" He kept the subject continually before him and waited till the mysteries of anatomy of the mosquito had begun revealing slowly by little and little into a full and clear light of message. The diligent thought on his 'black bodies' was the occasion of still further thinking, until at last he had brooded with the whole energy of his mind upon the subject – the oocysts on mosquito stomach wall. It was in Ross's case, as in every other, only by diligent application and perseverance that his great discovery was achieved. In fact, in achieving solution to the 'Great Malaria Problem' he had made a great service to the humanity and he gave full credit to his habit of functioning arduously and patiently: "*Fama mobilitate viget, vires acquirit eundo.*"

Dr Ronald Ross was born in an illustrious family, most of whom were either senior army officers including his father Major General C.C.G. Ross, serving in India and almost invariably all were highly educated with varied interests ranging from professorship, artists, musicians to golfing, mountaineering and alpinism. His birthplace was in India's Almora in the Himalaya but his early education, schooling and medical studies were completed in England. He developed a special penchant for poetry and mathematics which he nurtured all along his life. It is of little surprise to surmise then that the extraordinary power of accurate remembrance which he displayed on various occasions had been originally trained under the discipline of his father. After securing medical degree in 1881, against his wish but under father's pressure and personal economy necessity, he joined the Indian Medical Service to serve it until 1899 whereafter he took retirement from the post of Major Surgeon and settled in England. While in India he worked mainly on malaria and, with the help from Dr Patrick Manson, discovered *de novo* the mosquito-malaria connection for which, in conjunction with his demonstration of malaria cycle in birds, he was awarded the coved Nobel Prize in 1902. He was the first Indian who was awarded this prestigious prize. His discoveries improved understanding about the deadly malaria transmission among humans through mosquitoes and helped guide to devise effective methods for disease control, thereby saving millions from death annually.

"Woods are lovely, dark and deep,
But I have promises to keep,
And miles to before I sleep,
And miles to go before I sleep."

Experts all over the world are divided into two schools of thought in regard with his scientific prowess. Whether or not, therefore, he was a genius and self-made scientist, is a billion dollar question which embodies within its fold two extreme 'Characters" associated with his life and works. While, on one hand, it is well known with proper documentation that merely through the dint of hard work and focused application he had won the Nobel Prize, honoured with the Knighthood of Britain (though he was Scottish, however, he was throughout referred to England), The Government of India founded in his honour the Sir Ronald Ross Institute of Parasitology in Secundeabad (...this is the same hospital building where Dr Ross discovered mosquito-malaria cycle in 1897) and released a postal stamp bearing his picture in 1997 during centennial celebrations of his great medical discovery; in Great Britain a Ross Institute was founded in his honour and several other lofty awards, prizes, distinctions were showered on him post the Nobel Prize, however, on the other hand, much has been written, ironically without any supporting documentation, that he was short-tempered, irascible, quarrelsome, obstinate and greedy. He was blamed to have even sidelined his mentor, Patrick Manson, who helped him enormously throughout career and scientific experimentations. Often his becoming a scientist has been outrightly discredited, with a tinge of ridicule, in view of his perpetual and deeply rooted interests in arts, music, poetry, mathematics and like since childhood and that he had no inclination toward a medical profession and, therefore, he became a scientist by accident.

Dr. Ross's life and work are humankind's hopes for future triumphs and a glorious example to millions of young and dedicated vector-borne disease specialists who looked upon him as a strong role model. He possessed all those traits which help a man rise phenomenally in life. Medical arthropodologists/ entomologists, in particular, must emulate his example of sedulous industry, unflinching focus and inexorable resolve toward commitment.

1. *Disciplined family environment:*

As said earlier Ronald Ross was born in a highly illustrious family where inclination for higher education was prolific. His father, Major General C. C. G. Ross was an extremely sincere and disciplined military personnel, although at home he was always very pleasant and playful among children, ten in all.

General Ross was a gifted artist and played melodiously a number of music instruments. He was a great story teller and loved to recount stories from several classic and world famous books to his children, mostly during long walks, excursion tours or by their bedside in night. Therefore, all his children essentially developed these habits early in their lives. Ronald Ross, being the eldest among all siblings, had the advantage of experiencing company of parents and learning from them more. He has seen father writing poems and writing daily in his diary the happenings of the day. The famous maxim 'Like begets like' applies very much to the Rosses; the young Ronald followed every bit of his father's traits from an early age spent both in India and England. In the pink of his career Ross established himself as a prolific writer in manifold ways.

2. *Interests varia:*

Ronald Ross showed immense interest for art, writing verses and poetry, in particular, since much earlier school days. Ross evolved into a prolific and versatile writer capable to put different hues of nature and daily happenings in his creations. He habitually wrote poems on most of the important events in his life. In fact he deliberately intruded themes from the various happenings in his compositions, pleasant or otherwise. His poetic works gained him wide acclaim and they reflected his medical service, travelogue, philosophical and scientific thoughts. Many of his poems compositely collected in *Selected Poems* (1928) and *In Exile* (1931). Some of his notable books were *The Child of Ocean* (1899 and 1932), *The Revels of Orsera*, *The Spirit of Storm*, *Fables* and *Satires* (1930), *Lyra Modulatu* (1931), and five mathematical works (1929–1931). He also compiled an extensive account on his malariological explorations in *The Prevention of Malaria* (1910) and another *Studies on Malaria* (1928). He published his much talked about and reviewed autobiography *Memoirs - with a Full Account of the Great Malaria Problem and its Solution* (1923) based on his carefully preserved correspondence, telegrams, newspaper cuttings, drafts of published and unpublished material, and all manner of ephemera such as personal diaries and photographs - virtually everything about himself.

He was so obsessed with his verses and poetry that he left no opportunities, however trifle and resourceless, to pen his thoughts in a blitzkrieg; during recess at school, dinner table, excursion tours, on board a ship or a train etc. etc. – virtually anytime available to him. Later, when he was in I.M.S. and had plenty of leisure at his disposal, he composed his essays, dramatics, poems, novels and did mathematical exercises. He had seen to it that all his compositions of any kind are published for wide circulation. In case there was no publisher available or accepting his creation he had even dared to print privately his manuscripts from his own pocket. *Inter alia* he encountered difficulties due to frequent transfers to far off and non-malarious areas which interrupted his malariological investigations causing much consternation both to him and to Dr Manson in England. But, during these uncomfortable moments he found solace in his deep-rooted habit of composing poems, verses, novels, mathematical exercises. All his poems selected themes from his own day-to-day experiences and events. He found great relief with his writings which helped him getting rid of depression that dawned on him several times during his service and research career. His character of being a true polymath seeded into him a very strong trait of finding a solution for any problem from within for he had multifarious qualities to win over adversaries of scientific research, as he had himself once said *"Science is the Differential Calculus of the mind. Art the Integral Calculus; they may be beautiful when apart, but are greatest only when combined."*

3. *Voracious reader:*

Young Ronald Ross was in the habit of reading all kinds of literary masterpieces such as novels, poetries, dramatics, music and other kinds of arts and culture material. He had a tendency to visit bookshops everywhere he was either holidaying or working. Like most distinguished discoverers, inventors, artists, thinkers, and workers of all kinds, owe their success, in a great measure to their indefatigable industry and application, Ross too turned all things to gold – even time itself – by cultivating the habit of glancing at the biographies and golden works of great men.

4. *He was industrious:*

Dr Ross's greatest virtue, however, was his cheerful, patient and focused perseverance. He was a self-trained scientist. It is unimaginable even today, after more than a century following his great discoveries, that he (i) manufactured his own microscope to investigate malaria parasites, (ii) developed his own staining methodology for malaria parasites in blood, (iii) developed his own classification or 'categorization' of mosquitoes as 'A type' (Brown mosquitoes, possibly *Culex* spp.), 'B type' (Brindled mosquitoes, probably *Aedes/Stegomyia* spp.) and 'C type' (Dappled Winged mosquitoes, possibly *Anopheles* spp.) categories, (iv) in order to search the peregrination of malaria parasites outside human body he dissected cockroaches, fishes, various different mosquitoes such as for example *Culex* and *Aedes* species and what not! Although he finally met success with his 'dappled winged' mosquito, his labour with other non-malarious mosquitoes was not going to be unrewarded: *"No labour goes waste"*, (v) He drew umpteen times the human blood, both from his military hospital patients and the civilians, sometimes by giving monetary incentives (read bribery), for examination of malaria parasites so that he could feed his 'dappled winged' mosquitoes on them for seeing their future pathway in the mosquito body, which he ultimately proved in Auigust, 1897, and (vi) when he got no chance to continue his work on 'dappled winged' mosquitoes, he successfully demonstrated full mosquito-malaria parasite cycle in birds using this time the *Culex* mosquitoes. He was able to do it because he trained himself in that great aim. With none or virtually little progress at first, he repeated his experiments again and yet again in all directions of hope and expectation, and with great patience.

"Courage is the thing,
All is gone when this is gone."

5. *Resolute determination with focused attention:*

In 1895 when thirty eight years old Dr Ross went to the laboratory of Dr Patrick Manson to see Laveran's parasite, the crescents, Manson had, as per his volition, determined him as his student and sermonized in detail all about the tropical diseases then evolving as a new discipline altogether. He told him to consider that life cycles of filariasis worm (*Filaria bancrofti*) involving *Culex pipiens* and the Yellow Fever involving *Aedes aegypti* were already known and what remained to be answered among all deadly and/or debilitating diseases is malaria. He reminded Ross that India was the best place for investigating mosquito-malaria cycle and he must consider his tenure there in the I.M.S. as a great opportunity to cash upon. Ross agreed with him and assured to follow up his "Great Induction" theory. When Ross followed it step-by-stem on various different mosquitoes belonging to his 'A' (Brown, possibly *Cules* mosquitoes) and 'B' (brindled, possibly *Aedes/Stegomyia* mosquitoes) types, he did not succeed at all. It forced him to rethink about Manson's "Mosquito-Malaria Theory" which entailed that mosquito ingests the malaria parasites while drinking the blood from malariated human being. The parasite grows and changes form in the mosquito body and when the mosquito dies after laying eggs the parasites also are released in the dead pool of water. So his hypothesis spoke of infection to man only when he drinks the water of such pools where the parasite carrying mosquitoes had perished. By then Dr Ross had almost understood that Manson's theory did not hold water and he needed to see beyond the parapet.

"He who does well one work at a time,
does more than all."

It is not clear whether until 1995 he was aware of Dr Giovanni Battistuta Grassi's research involving anopheles mosquitoes, though still inconclusive. In those days communication was dead slow only through seaways between two continents, such as Europe (e.g., Italy where Grassi worked) and Asia (e.g., India where Ross worked) and many times letters had never arrived destination such as for example Dr Ross's first communication to Dr Manson written from the ship in 1895 had not arrived him, hence never acknowledged. The exchange of scientific literature was rather poor. In such scenarios of communication system then prevailing Dr Ross steered his research forward and worked on mosquitoes without knowing even elementary facts about the animals, leave aside their nomenclature and identification. Unfortunately, no literature on Indian mosquitoes was also existing when he was at mosquitoes' hunting. Despite all these inconsistencies, deficiencies and adversaries he did not give up his resolve and steered forward knowing very well all his efforts were actually on 'trial and failure' which in those times used to be the most dependable methodology. His basis microscope had rusted, eyepiece cracked, light reflector was damaged but he continued incessantly dissecting one mosquito after another, with all the microscopic observations properly noted for the day in his diary. Failure after failure befell him but he continued working undeterred through deteriorating vision and seething backache due to several hours of working with the microscope. He was improving with every unsuccessful outcome.

"Learning is excellence of wealth that none destroy,
To man nought else affords reality of joy"

Just when he thought the imaginative discovery was not his cup of tea the destiny smiled on him and handed him out the greatest of gifts - the sight of 'dappled winged mosquito', the key player in his entire endeavour to demonstrate mosquito-malaria life cycle. He laboured day and night, and sacrificed scores of similar mosquitoes with a phlebotomist's precision. He finally tasted the success on 19th August but because he wanted to be doubly sure he did not announce about this grand achievement. Next day, the 20th August he boldly confirmed having found the 'oocysts' with black pigment, confirming two important facts of the parasite's biology, *i.e.*, (i) The sexual bodies and/or crescents in the malariated human blood when arrived a mosquito's stomach change shape and size to sphericules, and after passing through some unknown stages of growth and development the parasite reaches the outer wall of the stomach to form specialized 'cells', now we know as oocysts. This whole thing may seem a simple affair to dissect out a malaria infected mosquito, yet what a long and laborious practice it requires in; twelve hours a day for ten years!

6. *Correspondence 'currente calamo':*

One of the finest characters that all scientific geniuses are said to have is to attend the correspondence promptly – on the day as they are received. Dr Ross's legendary prowess in this respect is something the young generation of scientists must emulate. He was an exceptionally gifted communicator – absolutely immaculate and meticulous in attending to a communication. Times and rigorous professional duties, rife with frequent and extended movements to new postings and/or special duties, had never seemed to hinder his energies and disciplined life in caring correspondence with his colleagues, particularly his mentor and guide Dr Manson, all over the world. Each of his letters is a speaking document of his day-to-day research work on malaria. It is noteworthy to note here that these were his letters with full scientific details to Dr Manson, in England, which the latter carried forward and deliberated on behalf of Dr Ross, in absentia, during important scientific meetings of learned communities in Great Britain. Therefore, when there was a tussle between Ross and Grassi over priority of research work to establish mosquito-malaria connection, due weightage was lent to these communications as a proof of work done much earlier than that of Grassi. The judgement of course went in favour of Ross.

Almost all of his letters to Dr Manson were written invariably in the late evening hours after returning from the day-long hard work on malaria patients, mosquito dissection or, more often, field work. Hardly influenced by the fatigue and loneliness he made a principle to report his work to Dr Manson on daily basis in the form of overseas letters even *currente calamo,* if unavoidable.

7. *He prepared scientific manuscripts for publication without delay:*

No doubt that an observation has to be replicated by repeated experiments before publishing into a formal scientific journal or read out in a formal scientific meeting. Yet, one must always bear in mind that there is always a way to understand the limits of such experimental repetitions. A delayed publication could become a secondary source of information and, however elaborate and irrefutable, it might lose the priority of publication. On this matter Ross was smart beyond comprehension, although he had certainly seen to it that his results were all reproducible and demonstrable time and again. He served British Indian Army where writing official diary on daily basis was ritually practiced. During his posting in Bangalore in early 1890s he had already earned a reputation for preparing excellent reports on combating cholera. Even he prepared four research manuscripts during his sojourn in Bangalore, ephemeral though. When it came to reporting on the discovery of oocysts in the mosquito stomach, he, besides having informed Dr Manson, in England, with almost all possible details through texts and drawings about his new find in August 1897, he also readily prepared a manuscript for an international scientific journal (*The British Medical Journal*) where it got published rather late on 18th December, 1897. Incidentally Grassi's publication, a more substantive with correct name of the vector mosquito, also appeared a few months later. Later when priority had to be decided between the two on the basis of dates of publication of the respective author's works, the decision, irrespective of the multiple methodological deficiencies galore in Ross's research publication, the priority was assigned to him, and not to Grassi whose publication was flawless, complete methodologically and zoologically well acceptable, on the basis of 'first' appearance of the publication – a big factor indeed in winning him the Nobel Prize in 1902.

"Time and tide wait for none"

8. *He created opportunities out of adversaries:*

All men get opportunities in their scientific professional life but not all can see them or capitalize on them. In fact, these opportunities hail so abruptly that only men with subtle alertness and scrutiny can catch hold of them to try out for extracting benefits.

"Fools walketh through forest and see no firewood,
Wise have their eyes in the head"

The long and glorious history of science is full of examples of several illustrious scientists who defied the deep scars of defeats but did not lose even for a moment their unflinching faith in their prowess of mind and body, and continued to work out their way to finally brace the success, even greater than their all the previous ones, and redeem the glory for an everlasting period. . "Slow and steady wins the race" is a well tested maxim true to all great successes in the history of science and research. No scientist can ever attain true success without having a strong character of patience and resilience: The world is his who has patience and industry.

It is said of Dr Ross, that when he had already finished a manuscript in a time-bound, correct manner, he would get up from his desk, go out for a walk, all the while filled in his head with thoughts about his manuscript, and after a few minutes' stroll come back to give a brilliant finish to his work that ranks amongst the best in the world! But, it will not do for everyone who would hope of writing an immaculate manuscript without properly assimilating beforehand all the essential data and develop a draft of the manuscript and review it time and again from all angles. The capability in putting these last vital touches is acquired only by the labour of a life, and the probability is, that the scientist who has not carefully trained himself, like the ever peripatetic Dr Ross did by going out in far and flung areas, away from home and family, would only fall short of finding a solution to his problem. Thus, a scientist has to move out of his airconditioned laboratory to seek the solution of his problem in the open, far reaching laboratory of the nature. He must understand the nature better and be aware of its infinite solution-offering potential.

चाह नहीं मैं सुरबाला के
गहनों में गूँथा जाऊँ,
चाह नहीं, प्रेमी–माला में
बिंध प्यारी को ललचाऊँ,
चाह नहीं, सम्राटों के शव पर
हे हरि, डाला जाऊँ,
चाह नहीं, देवों के सिर पर
चढ़ूँ भाग्य पर इठलाऊँ।
मुझे तोड़ लेना वनमाली!
उस पथ पर देना तुम फेंक,
मातृभूमि पर शीश चढ़ाने
जिस पथ जावें वीर अनेक

9. *Ross was ever-peripatetic:*

Being in the Indian Military Service Dr Ross was moved frequently to different places either on new posting or some assignments. He always followed instructions without a fuss or making some kind of representation. Even when he was deeply engrossed in his malariological explorations, sometime even while availing leave (e.g., in Sigur near Ootacamund he was on two months' leave but he employed self in his research work by daily marching many kilometers to Kalhutti, camping in forest rest house, all alone, with family [Two Daughters: Dorothy (*b.*1891) and Sylvia (*b.*1893), and one son: Ronald Campbell (*b.* 1895); the fourth child, a son, Charles Claye was born after Dr Ross was pensioned off in 1899 and had returned England, in 1901)] left behind in Ootacamund. It was in Sigur that he saw for the first time in his life a 'dappled winged mosquito', possibly an *Anopheles* mosquito.

"Those who lie in wait get
what others have left behind
after taking away the best
available in the lot."

There were umpteen happenings like this in his career when he was very near to his objective in 1897 but had had to wind up without prior notice and report elsewhere such as Kherwar in Rajasthan, a non-malarious area where there was hardly any possiblity to continue his work to further demonstrate the life cycle of the parasite in the mosquito body. Instead of complaining he silently endured all these difficulties, but of course he did use to mention about these professional problems to his mentor and guide, Dr Manson, and request him to use his lofty influence to get him posted to a malaria-prone area so that he could complete his work. He was truly an ever peripatetic scientist who converted adverse situations to his favour simply by dint of hard work and focused application.

Amazingly Ross was transferred or ordered to move for long duration on duty for over a dozen times in his rather short period of service in the I.M.A.:

1. Madras (Chennai)
2. Secunderabad
3. Bangalore
4. Ootacamund/Ooty
5. Bombay
6. Oodeypore/Udaipur
7. Kherwara/Nasirabad (Rajputana/Rajasthan)
8. Calcutta, and many other places nearby.
9. Punkabari, Darjeeling district
10. Kurseong
11. Nowong and Silchar (Assam)
12. Quetta
13. Myan Mar (Erstwhile Burma)
14. Andaman ils.

In conclusion, Dr Ronald Ross was one of those rare scientists who by their energetic character not only practically educate the budding researchers in habits of industry, but by the example of diligence and perseverance which they set before them, largely influence the scientific activity in all directions and contribute in a great degree to form the national character. The national progress, it is aptly said, is after all the sum of individual industry, energy and uprightness. His was unquestionably a mind of the very highest order. He was a thinking scientist – a genius signs of which he had begun to exhibit even before setting out on his first first furlough to England in 1888, to prepare for a Himalayan cause ahead.

●●●

5

Preparing for a Himalayan cause: First Furlough to England (1888-1889) and Back in India – *First Major Encounter with Mosquitoes* (1890-1893)

"It is not in the most distinguished exploits that man's virtues or vices may be best discovered, but frequently an action of small note, a short saying, or a jest, that distinguishes a person's real character more than the greatest battles or the most important actions"

— (Samuel Smiles in 'Duty', 1880)

Dr Ross left his sea voyage at Malta and travelled onward by himself through Sicily, Italy, Switzerland, and France – countries which he had never visited. He praised the beauty of their cities and civilization compared to India's bout which he ignorantly used words like 'savagery' and 'exile from India'. The young and inexperienced mind of Ross was unable to judge the standards for measuring the beauty of a country. What he noticed in Rome, Naples, Florence and Venice was physical or external beauty compared to the much older and rich civilization in the lands of mighty Sindhu, Ganga, Yamuna, Brahmaputra, Cauvery and Narbada etc. rivers (Fig. 5.1). Also, it was the state of mind, fearful and home-sick, which led him to then think 'on exile from India' on his seeing another white brethren. It is, however, an irony' that, first, the same country, India, whom he cursed, in course of time blessed him with the world's highest scientific decoration, the Nobel Prize followed by innumerable distinctions and recognitions, secondly, Dr Ross who, in his disputed claim for priority of his discovery of mosquito-malaria connection, got entangled in a furious bickering with only the Italians; thirdly, Dr Ross was blamed for indirectly using his high acquaintances in United States of America, by many of his 'white' only colleagues and friends to have stooped so low by Germany etc. and putting influence on Dr Robert Koch, the independent arbitrator appointed by the international tribunal, to settle the Ross-Grassi spat in his favour. , and lastly, Dr Ross was spared by an almost certain embarrassment at the hands of his very sincere, honest and disciplined Indian laboratory assistant colleague in Kolkata, Kishori Mohan Bandhopadhyay. It is no exaggeration that he worked all important works in Ross's laboratory and, with a little bit of British Government's honest weightage, could have even been jointly considered, along with Dr Ross, for the Nobel Prize! Life's experiences teach that a scientist speak out on his results only after confirming them through repetitive experiments for their confirmation and Dr Ross was still learning in this new school of science and research. His views on India were soon going to be changed for this country gave him what he could never have received anywhere in the world, the Nobel Prize, loyal support and and help in pursuing research and, above all, love of her people..

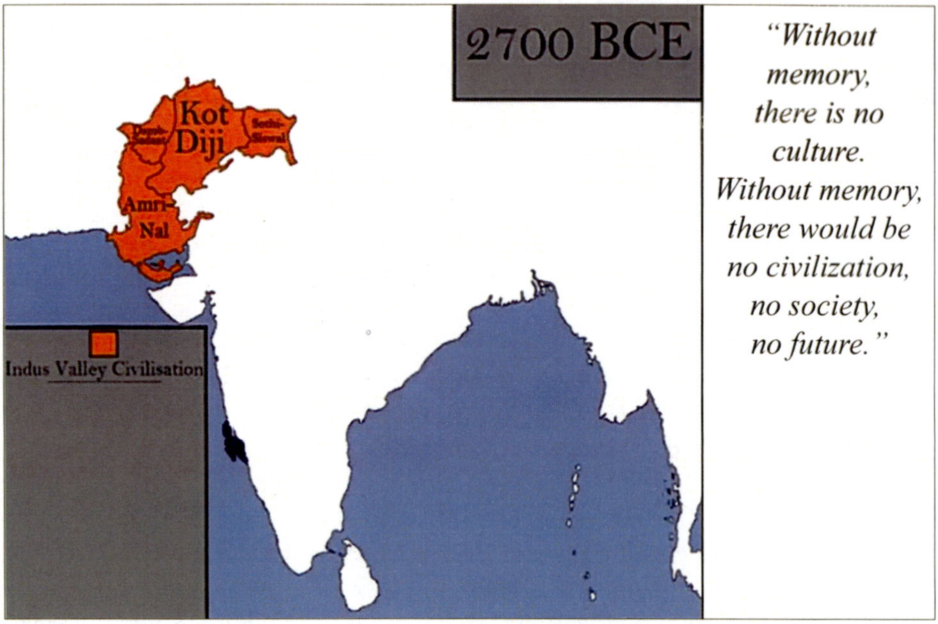

Fig. 5.1. Depiction of Indus Valley Civilization (2700 BCE) – one of the oldest and richest civilizations in the world.

Dr Ross arrived Newhaven (England) via Paris (France) on a train. He suddenly found his old spirit bounding within him, His delight knew no bounds on meeting with his family living in England; parents, sisters and brothers. His brother Claye, who had also joined Indian army and was attached with the 14th Sikh regiment, was also at home leave. His other brother Charles had only a few months ago entered the British Indian Army, while another younger brother John was busy preparing to join the army.

Externally though Dr Ross was taking interest and giving a hand in whatever type of work his parents or others in home asked for him to attend to, but internally he was actually wresting in his mind with his own diverse thoughts and struggled to primarily determine his future course of action in his professional duties back in India. He had seen much of rural India during his long inland and sea journeys, and his short yet significant exposure to cholera and dysentery-hit Bangalore civilians, re-enforced his thinking that India's biggest challenge and problem was sanitation that occurred abysmally everywhere, be it a town or village; flies, mosquitoes, cockroaches etc. were everywhere. Water supply, even in advanced city like Bangalore, was faulty; in most places the water conduits ran close to sever lines, with likelihood of picking up infective organisms from the latter; this is how he concluded the possibility of emergence of cholera outbreak in Bangalore a few years ago! He was already very impressed by the discoveries of Pasteur and Koch, and the newly emerging science bacteriology was really coming to the fore in a big way. Therefore, he decided to specialize in bacteriology to equip his knowledge on the microbes. He proceeded first to take admission in Diploma in Public Health at the Royal Colleges of Physicians and Surgeons, with Professor W.R. Smith, and later in microbiology under Professor Klein, F.R.S., in London, passing examination in December, 1888. It is noteworthy to mention here that he passed the examination due simply to his excellently solving a difficult question on population through his own self-educated mathematical ingenuity. This implies that nothing in this life genuinely acquired goes waste and often at times of test it comes quite handy to accomplish a task in time when others could not.

Dr Ross tied up nuptial knot with beautiful Miss Rosa Bloxam on 2nd April, 1889 and he found in her a very dependable life partner. He felt very energized and a great change from his solitary living in India. At the same time, however, he felt obliged to the Indian government for having granted extension of his leave for two months, to study bacteriology incorporating pathologic microbes, microscopic investigation, culturing microbes etc. He drudged his way to pass the examination, sacrificing a lot of his personal life's time, but still the word malaria was a distant imagination albeit the clouds of skepticism becoming thinner and gradually fading away fast. He resolved to keep *honest* in his investigations through all the thick and thin.

> *"God give me unclouded eyes and freedom from haste.*
> *God give me a quiet and relentless anger against all pretence and all pretentious work and all work left slack and unfinished.*
> *God give me a restlessness whereby I may neither sleep nor accept praise till my observed results equal my calculated results or in pious glee I discover and assault my error.*
> *God give me strength not to trust God!"*

On 1st August, 1889, Dr Ross accompanied by his pretty wife set on the sail to India through a very hot passage of the Red Sea, and Colombo, arriving Madras on 6th September, 1889 where, before he could relieve himself from the overweighing luggage, he was handed out his relieving orders to leave on field duty to Myanmar as Medical in-Charge of the 2nd Regiment Madras Infantry. He was obviously displeased, looking at his situation, but conformed to the orders since he had no other option available to him. With his wife in accompaniment, for the first time away from her home in England, he for a moment was terribly irritated. His consort, Rosa, coolly advised him to not think so worrisomely and that some could be certainly found out soon. Ross looked at her with wonderment reasonable solution and a sense of gratitude for her quiet way of dealing with the subject. He made an arrangement for his wife to stay back by herself in a hotel in Madras and left for his destination on 21st September. His wife, too, joined shortly after, but she was mostly alone there as well without her husband who was ordered to march out farther. Here, after living for nearly six months, he received orders to rejoin at Madras. By now Ross has won many hearts in his regiment and was certainly in good books of his superiors. Accompanied by his wife, who rejoined him *en route*, they left Rangoon (Yangon in Myanmar) for India on 29th April, 1890 – to join this time as the Staff-Surgeon, a much lofty rank and his first permanent appointment, at Bangalore, his old temporary post. He wondered if it was a reward for his hard work and dedication? At least his wife thought so.

In Burma, he encountered a lot of mosquito breeding which attracted his attention beyond observation from distance. He collected larvae, bred them from larvae. Removed them from his residence and was able to distinguish two main types which he called 'grey' [possibly a Culex] and 'brindled' [possibly a Stegomyia (= Aedes)] mosquitoes, respectively. He searched at M/s. Higginbotham and other bookshops in Madras but could not get any book on the subject of mosquito identification. Even though Linnaeus (1762) had already described, amongst many mosquito species, a medically important mosquito, *Aedes aegypti*, from a non-Indian specimen, the mosquito was nevertheless prevailing abundantly allover India. Dr Ross's limitations with the mosquito biology and identification were understandable; he was not a zoologist and he might certainly have not studied Linnaeus's *Systema Naturae* (1758) and its subsequent editions particularly the 10th version. He seemed to have also never known about a dangerous mosquito such as *Anopheles claviger* (Meigen, 1804), common in Italy, Finland and possibly some other European countries, *Anopheles vanus* Walker 1859, common in Indonesia, or even *Anopheles barbirostris* Van der Wulp (1884) prevalent in India. Thus, he was very elementary in his knowledge about mosquitoes on whom he had so far not read a single research paper or a book. This was largely because he was obsessed with his literary compositions, arts, music and mathematical exercises which apparently did not connect gainfully with his routine medical services in military terms. His mind was once again engrossed all the time by the same old scheme of work – composing poetry, assays and novels. He was sailing on two boats! Mosquitoes had not occupied his attention yet!

After waiting for some weeks in Madras Dr Ross and his wife left for Bangalore by train where they arrived on 20th May, 1890. This was his wife's maiden journey to Bangalore and she admired the landscapes. He took over charge of the coveted post of Staff Surgeon for three years and was commensurately quartered in a delightful bungalow within the premises of the garrison. He soon found out that the water supply and storage means were grossly faulty and, consequently, a large section of people from the officers and their families always complained of enteric or other milder fevers and bowel disturbance with parasitic infection. He observed presence of malaria among few patients but this was invariably introduced from outside. He suddenly felt a surge of dutifulness and determined to work really hard at his profession. Like in past, he had time aplenty to devote to all kinds of medical books and research journals that he could lay his hands on in Bangalore. With an almost rock-like grit determination he resolved to take full interest in his profession and give his best shot there.

"If you salute your duty,
You no need to salute anybody,
But if you pollute your duty,
You have to salute everybody."
– (Dr A.P.J. Abdul Kalam)

As the Staff Surgeon and medical officer he got extremely worried about the ongoing fevers, particularly among the children. These were characteristically mild fevers and invariably all suffered with intestinal affections. Such a fever, he conjectured, could be malaria and related the fever with intestine. He later found out that his guessing was utterly wrong and all those fevers were actually typhoid or of some viral origin. He knew about Laveran's discovery of malaria parasite in 1880 but was not entirely convinced the malarial parasite is present in human blood and presence of which could be demonstrated under a microscope. On the other hand Dr Ross tried to emphasize that what people were saying were malaria parasites did actually represent the artefacts in the coloured stain. Also, he rejected completely the 'miasma' theory for malaria transmission. He was correct in the latter case but absolutely mistaken in the former case; his thesis that artefacts mimicked malaria parasites did not hold water at all. He later corrected himself as his own tryst with the disease was fast approaching nearby. In 1901 he, and one of his fellow officers, Mr Tait, went on a fishing tour in the Bhawani river close to Metapolliam at the base of Nilgiri Hills. They treaded at a stretch up to Coonoor (5000 ft. MSL), about 36 Km from Metapolliam. Following this expedition Dr Ross came to know that malaria gripped Mr Tait at the rest house near Metapolliam; he was himself saved because he used the mosquito net on his bed before sleeping. He gauzed over this difference but could not implicate the unsuspecting mosquitoes as a reason behind the happening.

Dr Ross stayed in Bangalore for three years, filled with the family-joy, sports, fishing explorations, literature accomplishments, and mathematical exercises – all but scanty medical and hospital services. But, in spite of these pleasures he began to feel depressed again as he received by a telegram the news of the demise of his father (20th June, 1892). He dashed to Murree, a mountain town (2291 meters), located in the Galyat region of the Pir Pinjal Range of the Himalayas. For he was internally dissatisfied he obliged himself to ill-find excuses in either the climate of salubrious Bangalore or with his on fitment to the surroundings being a 'white'. He became oppressed to such an extent that he had planned to earn a compulsory retirement after his first pension became due in 1897 (the same year when he made the pathbreaking mosquito discovery at Secunderabad!). Therefore, as soon as his appointment as Staff-Surgeon had ended in April, 1893, he availed leave for three months, staying in Coonoor, near Ootacamund – to rest and for a long due recuperation, and of course to recommence the incomplete verses.

About June, 1893 Dr Ross left Coonoor for Secunderabad, passing on way through Bangalore. The hot and sundry journey was completeled in about a week. In Secunderabad he had to report as a temporary medical officer in the 20th Madras Infantry. Maintaining his stoic gesture throughout, Dr Ross, now a major, was internally feeling extremely vexed and humiliated at the hands of his government as he was not only 'let down' from his previous elevated rank, he was still lingering on an '*ad hoc*' appointment and obliged to incur monetary loss also of Rs 100 a month (which in Ross's mind was staggering at an estimate of Rs. 100 x 12 years = 14,400). He was not happy with his quarter, too. The only recourse, of course, was the climate which suited him better than in Bangalore.

"Problems cannot be solved
at the same level of awareness
that created them."
— (Albert Einstein)

● ● ●

6

Decision on Malaria: *Second Furlough and Dr Ross's meeting with Dr Patrick Manson – his Mentor and Guide* (1894-1895)

"Life is a matter of choices,
and every choice you make makes you"
— (John C. Maxwell)

Secunderabad, one of the famous twin city located about 14 Km east of Hyderabad – the other and older city, was established as a large military cantonment area where Dr Ross made his laboratory-cum-office in a soberly looking high roofed and large sized three-room building in Begumpet, with his residence at a stone's throw. The landscape interspersed by several little lakes and the surroundings during monsoon season appeared all quite beautiful, verdure and sparkling. Malaria and mosquitoes, too, abundantly prevailed. It was great opportunity for Dr Ross to equip himself in the science of practical malariology, albeit his complete ignorance about mosquito's role in the disease transmission nor did he ever even had the opportunity to notice a malaria parasite. So he resolutely set to find Laveran's malaria parasites in the blood taken from the pricked fingers of symptomatic patients, but in vain for he followed an incorrect methodology - of his own! He repeated his efforts even resorted to cautiously drawing of blood for his examination during both day and night times – leaving no doubt in his mind about the procedure followed, but of no success yet. He was sorely disappointed by the absence of favourable results, on one hand, and exceedingly frustrated due to the feeling of lack of sufficient microbiological knowledge acquired, on the other. His malariological endeavours were soon disrupted when he was ordered, to his great delight though, to move out to a permanent appointment as surgeon to the 19th Madras Infantry at Berhampur in Bengal. He, joined by his wife and two little kids, had a tying journey throughout which mosquitoes had also mercilessly feasted on them. In Berhampur, a charming little military settlement, he spent a rather uneventful happy time much engrossed in his regimental duties. However, since for past four years neither he or his wife had been to home in England and, more so, nor his children had thus far ever been there, but also the nagging question of never seeing a malaria parasite by himself, he was obliged to avail his second furlough in England in February, 1894.

Throughout journey, which lasted about a month, he almost regrettably thought again and again about reasons of his failure to notice a malaria parasite in the finger-pricked blood. He had searched numerous human blood specimens, but due to sheer bad timing and/or circumstance, he did not make headway. All his preparations were negative for *Plasmodium* species at all. Consequently, he felt that all his hard work to yield malaria parasites in symptomatic human blood during 1990-1994 had gone waste, He nonetheless did not lose heart and stuck to his gun firmly. He thoroughly reviewed his methodology and quickly succeeded in rectifying the errors involved. Failures educated him and he became determined to become a competent microscopist. Arrived in England, he immediately began to find a solution to his gnawing problem, the malaria parasites, but continued to have diversion of his mind by his other comprehensible regular preoccupations – the stories, poetry etc.

By 1994 Dr Ross had access to worldwide important books and other literature including scientific journals and publications. To many of the journals he subscribed and much of the literature had arrived him through intense and regular correspondence with the concerned scientists across the globe. He understood how much were all the nations living in tropical and subtropical regions of the world affected with malaria, and India, along with ancient Greek, where had been quite many scientists who shed doubt if mosquitoes, who were copiously prevalent in marshy areas where malaria, justifying the nomenclature (mal=bad; and aria=air), virtually persistently occurred, had anything to do with the disease? He also understood that the whole world was gradually getting acquainted with the fact of severity caused by malaria to health, economy and intelligentsia. He further acknowledged that "in Italy especially where of all civilized countries* the disease was most

---

* Here it is noteworthy to mention that Ross was completely ignorant as to who was to be called a 'civilized country' the one which possessed all the resources and its people happily lived in harmony through ages, such as for example India, or the one, like England, who under the garb of commerce through East India Co., first pleaded to the Indian kings to allow them to initiate trade, and then, on surging their greed, attacked India's sovereignty by all kinds of treachery, tyranny and the 'divide and rule' tactics, plundering all her resources and impoverishing its people and more, and eventually siphoning varied treasures over to the 'Great Britain'. Britishers brutalized in every possible way the India's nonviolent movement for complete freedom which they attained on 15th August, 1947 by throwing out the tyrant rule of the Britishers once for all! A dishonest, untrustworthy and brute nation can not be a 'civilized country'. India, with its glorious past and immaculate history, has always been a hardworking and harmonious country who never ever attacked any other nation or overthrew their rules to dominate them. India certainly is a civilized nation.

prevalent, this process of understanding the disease epidemiology was most apparent" (Ross, 1923).

The Italians had floated the hypothesis of *paludic miasma* based on the observation that malaria abounded in areas with marshes and that it all concerned some kind of infective emanation, either chemical or organic, from the heavily vegetated stagnant water. The hypothesis could never be scientifically founded since a confounding observation was that malaria occurred even in areas where no marshes were present. This innovation led to *telluric miasma* which meant that poison emanating from the marshes had its cradle in the soil. The speculation did not hold here and even associated the disease with the vegetation growing in marshes, with fungi or bacteria which enter the blood causing fever. All these were discredited in course of time.

An altogether different speculation was put forth by the Italian Rasori, in 1846, that the intermittent fevers were result of the parasite's reproduction but even this extraordinary prediction was dismembered in the want of evidence. Meanwhile H. Meckel in 1847 came up with his conclusion that black pigments, now known as haemozoin, had a connection to characterize malaria. Recent advancement in malariology has shown that haemozoin is actually a disposal product formed from the digestion of blood haemoglobin by the malaria parasite. Almost three decades later the output of C. Gerhardt's research had discredited this conjecture on the ground that a healthy man could be malariated by the blood of an infected person.

The most powerful and convincing discovery was put forth by Dr A. Laveran, A French army surgeon on duty to Algeria where, on following the pathway of black pigments, in patients' blood from his armed core he found what is today recognized as gametocytes of *Plasmodium falciparum* and several other stages of development of another parasite *P. malariae* (Laveran, 1880; Fig. 6.1a). He simultaneously discovered that in the infected red blood corpuscles there were invariably present one or few black pigments which grow in size filling the corpuscle almost entirely. On bursting these infect fresh corpuscles. Besides, Laveran observed such black pigments with an amoeboid movement in the serum as well where the incredibly wonderful exflagellation phenomenon happens to these bodies – a discovery that revolutionized medical science in the tropics. Subsequently in 1886, C. Golgi, in Palvia, Italy, demonstrated biology of the malarial parasite as well as quartan (in case of *P. malariae*) and tertiary (in case of *P. falciparum*; to this list is later added *P. vivax, P. ovale* and, most recently also, *P. knowlesi*). Some time later Canalis, Marchiafava and Celli confirmed the biology of *P. falciparum.* Viewing the malaria parasites under microscopes with distinct and well-defined cytoplasmic and nuclear contours always remained a challenge and very serious drawback in studying morphology of the parasites. This problem was solved when

Ramanowsky found the proper way to stain all the parasites with equal clarity. Many scientists across the world, however, failed to figure out the exact future of the sexual forms of the malaria parasite. That is, how these parasites manage to get an entry or appear suddenly in the human blood. As mentioned aforesaid many theories were floated but none was sound enough to be convincing. As we would later see it was here where the role of Dr Ross exactly fitted. It was not clear as to whether he knew then about the Italian genius Giovanni Battista Grassi's malariological advances or also the great feats achieved to control Yellow Fever in Central and South Americas due to epoch making researches of Dr Finlay and Dr Walter Reed (Figs. 6.2a,b; 6.3a)). Clearly he nevertheless found a humongous vacuum of information awaiting to be explored in malaria, in particular.

"No research is ever complete."

It is not accident, then, that helps a man in the world so much as purpose and persistent industry. To the feeble, the sluggish and purposeless, the happiest accident avail nothing, - they pass them by, seeing no meaning in them. But it is astonishing how much can be accomplished if we are prompt to seize and improve the opportunities for action and effort which are constantly presenting themselves. Dr Ross had a keen eye for everything malarious; the subtlety with which he grasped the matter was astonishing. Thus, the most significant turning point in his entire career occurred when he was directed in March 1894 by Professor Kanthack, a pathohologist at the Bartholomew's Hospital in England, to consult Dr Patrick Manson, also living in England (Fig. 6.1b). Prof. Kanthack knew very well about Dr Manson's grand explorations in tropical diseases and also about his great filariasis adventures and discoveries. He had sensed a boundless anxiety and desperation in Dr Ross's conversation, and, therefore, he felt that a seasoned physician and scientist such as for example, Dr Patrick Manson, with enormous exposure to arthropod-borne diseases, only could help to clear his doubts, whatso ever. Dr Ross (Fig. 6.3b) wrote a very polite and brief letter to Dr Patrick Manson, seeking an appointment on 10th April, 1894.

(a)

(b)

Fig. 6.1. (a) Dr. A. Laveran, and (b) Dr Patrick Manson

(a)

(b)

Fig. 6.2 (a) Dr Walter Reed, and (b) Dr Carlos Finlay

(a)

(b)

Fig. 6.3. (a) Dr Giovanni Batistuta Grassi, and (b) Dr Ronald Ross.

गुरु बिन ज्ञान न ऊपजे, गुरु बिन मिले न मोक्ष ।
गुरु बिन लिखे न सत्य को, गुरु बिन मिटे न दोष ।।

(Without the Guru no one obtains spiritual knowledge
or achieves salvation.
Without the Guru no one can see Truth
or have his doubts removed.)

Dr Manson, a tall and handsome person approximating fifty years, had then emeritated from China for some years and was having a very flourishing private practice in London. He received Dr Ross with warm welcome at his residence at Queen Anne Street, Cavendish Square. During this maiden summit Ross could only detail out about various of his experiences in Indian military regimental hospitals, with only passing reference to malaria. Dr Manson was very kind and gave him all ears and took much interest in his narratives. Ross subsequently met him many times and quickly converged on the objective of his rendezvous; to know more about the disease of malaria, particularly Laveran's malaria parasites and the all-important question as to what happens eventually to the sexual forms, the micro- (male) and macrogametes (female), being the last of the stages in parasitic development in the human blood? In order to understand Ross's viewpoints on tropical diseases, particularly malaria, Dr Manson as in vogue for him all along his past often took him on long evening walks and carefully developed a list of queries out of their conversation. For Ross was dying to see Laveran's malaria parasites, Dr Manson first of all graced him with the microscopic slides containing Laveran's bodies or the 'crescents', belonging to *P. falciparum.* He saw the parasites again and again till he was completely satisfied. Dr Manson could mark a child-like delight on the face of Ross who was speechless with gratitude. Dr Manson, looking at Ross's face, has softly but cautiously asked him if he had any doubts? Ross, moving his head left and right, simply beamed forward a bit in gratitude and murmured, "Thank you. None". Manson also took him to Charing Cross Hospital, where he showed him other forms of *Plasmodium* and, also on several occasions, with great kindness, to the Seamen's Hospital in East London, where he acquainted Ross with some of the most important and rare literature and books by the world's famous malariologists of that time such as for example Dr Mannaberg, Dr Marchiafava and Dr Bignami. His doubts which had been nagging him for so long had vaporized, giving way to pure thoughts alluding to his future roadmap on his experimentations. For the last time, thus, Ross put up to Dr Manson his one major doubt, "Though it was certain that malarial fever is caused by parasites in the blood, and the disease had actually been communicated from man to man by the inoculation of the blood of infected persons nobody knew as to how these parasites had managed to pass from the sick to the healthy *under natural conditions.*" Ross called this vacuum in knowledge the Great Problem of Malaria. Dr Manson was impressed by his quick and clear comprehension of the whole equation of life cycle of the parasite and the unanswered questions on the existing gaps in knowledge. He recounted to Ross how some scholars or savants in recent past had casually referred to mosquitoes as the carrier for malaria but their broadcast had thus far remained entirely unfounded in the absence of any scientific experiments. Manson patiently told him his own story about the discovery of *Filaria bancrofti* (=*Wuchereria bancrofti*) worms in the lymphatics and how the young ones of the parasite are sucked up with the blood by mosquitoes (possibly a Culex species such as *Culex pipiens*) in which they grew and developed to a considerable size (Manson, 1879). He conjectured (incorrectly though, as we later knew of course that the parasite, still in the fully grown larval stage (L_3) in mosquito body, is actually passed to a healthy person during subsequent bites for blood feeding, through food canal passage) that parasite or filaria embryos come to water after the infected mosquito had oviposited there, wherefrom they entered human body after they had drunk the parasite-infected water. Dr Ross quietly heard the whole story but surely many questions surfaced in his mind because he had felt there were too many gaps in his narrative.

Dr Manson treated him with the greatest kindness and generosity, and Dr Ross knew from him all he had wanted to tell. One day, during their sojourn at the Seamen's Hospital, as they were walking along Oxford Street at about 2.30 pm, Dr Manson, then convinced that there was little left for anybody to find out in filariasis, drew Dr Ross's attention to his 'Mosquito Theory' for malaria. Dr Ross nodded in conformity since he was already aware of his conjecture in the book written by Laveran titled "*Traitè des Fièvres Palustres. Doin, Paris*" (1884). Dr Manson explained, "We knew the function of the asexual forms of the parasites, but were ignorant of the functions of the sexual forms or crescents and flagellate bodies." He went on, " Shortly after blood containing these forms is drawn from a patient's finger the crescents swell up, become oval, and then spherical, and lastly sometimes emit the long and actively motile filaments." He continued in the same breath, "These filaments may sometimes be seen under the microscope to break away from the parent cell to wriggle about rapidly in the fluid of the blood." These free living bodies were the big question of the day for all those who researched on malaria. While one school of thought (Italian) opined this represented the struggles of the dying parasite while other school of thought (Laveran and Danilewsky) held that they were living entities and even could connect to an extra-human body life history (Mannaberg). Dr Manson floated his cogent interpretation, "When the mosquito sucks the blood, these sexual forms enter its stomach, and there, in a few minutes, emit their motile filaments just as they do under the microscope." He concluded, "The motile filaments were flagellated spores which next pass through the walls of the mosquito's stomach and take up their abode in the tissues, where they must develop into some unknown stage and then enter drinking water." At the end of the narration, Dr Manson looked up on Dr Ross's face who was of course immensely impressed with his argument. He replied enthusiastically, "I will follow up to test the hypothesis thoroughly on my return to India." However, Dr Ross was still ignorant for malaria of many latest scientific researches and, in fact, he plunged into the fray without getting adequately equipped scientifically; one of which, of course, was his blatant ignorance about the types of mosquito and their identification. He was destined to study the *Plasmodia*, not in man, but in the mosquito – a thing which had never been attempted at all! India was awaiting his best shot at malaria parasites yet!

Dr Ross had an enjoyable time in England, also in Switzerland on a pleasure trip along with his and a friend's families. Close to the end of his leave, he was also immensely saddened by the untimely demise of his dear brother Claye, four years younger to him, who was Killed in action near Chitral in India. The gruesome massacre of the army men at the hands of treacherous ambushing tribesmen was narrated by the sole survivor of this battle in Chitral near Gilgit in Kashmir, Ghanda Singh who escaped miraculously after several days solitude and hunger – to tell the tale.

Dr Ross's summit with Dr Manson had many ramifications. In Dr Manson, Dr Ross had found not only a great mentor but a father-like sincere guide and friend. They developed later an important bond between them, and the long-time and diverse correspondence spanning two decades became an invaluable treasure in science. With several new ideas and facts, and his newly invented microscope, in his weaponry, nonetheless, he left England for India on 28th March, 1895 – for most eventful four years of his life - embedded with ceaseless industry, unflinching focus and resolute application. Dr Ross was all set to solve the 'Great Problem' of malaria - a grand piece of research which the life of millions in the world.

"Who comprehends his trust, and to the same
Keeps faithful with a singleness of aim;
And therefore does not stoop, nor lie in wait
For wealth, or honour, or for worldly state;
Whom they must follow, on whose head must fall,
Like showers of manna, if they come at all."

— (William Wordsworth)

This time Dr Ross departed for India, all alone by himself, leaving behind at home his wife and three children; two daughters and a son.

• • •

7

Back to India: The First Practical Lessons in Malariology – *Ambushing the Beast in Mosquito* (1895)

"Take up one idea. Make that one idea your life – think of it,
dream of it, live on that idea. Let the brain, muscles,
nerves, every part of your body be full of that idea,
and just leave every other idea alone.
This is the way to success."

— **(Swami Vivekananda)**

Leaving England for India, Dr Ross boarded *Balaarat*, a steamship, A slower sailing by the steamship offered much wanted time to Ross. He availed the opportunity to immediately set to work on board itself. He lectured to all passengers on ship and explained about deadly malaria disease. He took them into confidence and made them to agree to sacrifice some blood off their fingers for his investigations. So, he commenced at once to prick passengers' fingers for examination under microscope, much to the donors' amusement since they were also looking at their own blood cells possibly for the first time in life under a microscope (Fig. 7.1)! When the ship had a brief lay off at Malta he rushed maddeningly to the Civil and Naval Hospitals, respectively, to examine the blood of a few feverish locals, but none was found to be positive. He was disappointed, but not down in spirit. He caught cockroaches on the ship and dissected them because he had read that these insects possessed some parasite like malaria organisms. He also tried to see the malaria parasites even in a flying fish. Alas! Success still evaded him. All along the microscopic works, however, he was vigorously consulting a large number of books and other literature on malaria, and even some on mosquito. He continued his voyage along the coast of Italy to Brindisi, a port city on the Adriatic Sea, in southern Italy's Apulia region, then to Greece, which he was seeing for the first time, and finally, via Red Sea, and after a transshipping to *Thames*, back to India's port city Bombay on 21st April, 1895. From Bombay he moved to his posting in 19th Madras Infantry at Secunderabad's cantonment called Begumpet. But, because he was stranded in Bombay for a couple of days, between 21st and 23rd April, he availed the opportunity in disguise again for examining malaria cases in the Civil Hospital; he succeeded to find under microscope parasites in the blood of five out of eleven patients. Then he turned his attention to mosquitoes, but the heat was awful and the wind like the blast of an engine-furnace. Still he was able to catch three mosquitoes which he put into a bottle. Since he was much impressed by the discovery of the first malaria parasite by Laveran, he was also keeping an eye on likely encountering one new species somewhere, particularly in India where people suffered due to malaria all round the year. On reaching Secunderabad he found the town experiencing height of unwelcome hot weather, with ground completely desertic, and wind hot and dry.

Fig. 7.1. Microscope assembled by Dr Ronald Ross for his malariological research. (*Source:* SRRIP archive)

Ross had weighed all the possibilities before focusing on mosquitoes for his discovery of the *Plasmodiua.* Although, Manson's teachings were greatly helpful in understanding mosquito's role in transmitting the parasite from an infected one to another (albeit an altogether wrong interpretation), it did not seem to much impress him and he resolved to investigate everything *de novo.* In order to achieve his motto, he needed (i) first of all, sufficiently large number of malaria (*P. falciparum*) patients with numerous sexual bodies and forms in their blood, (ii) secondly, numerous lab-cultured mosquitoes from larvae, and (iii) thirdly, such lab-emerged mosquitoes must bite and feed on the malaria patients.

At Begumpet Ross's hospital contained two wards, besides offices in a single-storied building. All his work on malaria and mosquitoes was carried out in addition to his duties as regimental surgeon and he had to bear all his expenses by himself.

Fig. 7.2. Dr Ronald Ross with his own fabrication of a microscope. (*Source:* SRRIP archive).

By now Dr Ross had developed enough expertise to study and distinguish the malarial parasites under the microscope within few seconds and also breed mosquitoes in his moderate laboratory. He studied more deeply the stained sexual bodies, the crescents, and the process of exflagellation, besides the asexual forms in the blood. He had the firm knowledge that these sexual bodies did not occur in a healthy person and were a characteristic only of a malaria-struck human being. With the pace of his achievements he grew really wild with excitement, save for the fact that the mosquito reared in the laboratory had not liked to bite, even himself as bait! He wondered if temperature of his laboratory had anything to inhibit mosquitoes to feast on baited human blood? More so the malaria patients in the Civil Hospital who volunteered to finger-prick somewhat grew skeptical of his intentions and continuously feared for the next action. Dr Ross spoke to them and, in order to win their favour, tipped them each with a rupee a prick, but just when he wanted to prick their fingers he found two of the patients had escaped without any word. He then turned his attention to the regimental hospital and on the sepoy case, but of no avail due largely to the fact that all army personnel were obliged to take quinine, an antimalarial. He did not give up yet and continued his efforts and was soon rewarded with a case who had suffered due to dysentery but his blood contained the crescents; next day the number grew many times more! He readily transferred the patient to his hospital and let lose mosquitoes on him to blood-feed, but of no avail. Mosquitoes did not bite at all, and for a moment gross disappointment enveloped him. He thought that there had to be a way out to persuade laboratory-emerged mosquito bite and feed on blood. He was not ready to accept defeat instead preferred to move forward with undeterred resolution.

We learn in our own life that failure are in fact the very basis for future success. A hardworking man always succeeds if he genuinely pursues his goal. The genius is in the habit of perseverance. Fortune has often been blamed for her blindness, though the fact is it is man who is blindingly ignorant of the inexorable laws of life. Hard times have always been best masters for scientists. They do not turn their back to difficulties, rather go ahead and tackle the problem head-on. Scarcity of help is a misfortune to all, not a scientist for he seeks no ease and riches, which, it is rightly said, are anyway not necessary for man's highest culture. Indeed, adversity induces in a scientist that indomitable spirit of vigorous self-help and may be converted even into a blessing, rousing him to that struggle with the adversary against which while even the most blessed financially may kneel down he would rise to the occasion and find in that the indubitable source of strength, confidence and triumph. Dr Ross was an apostle of perseverance.

Dr Ross searched for literature on mosquitoes primarily with a view to comprehend as to which kind of exact mosquitoes were concerned in malaria.

He tried everything possible in those days of scarce bookshops and snail-mails to lay his hands on useful literature but he was not even able to obtain any scientific literature on Indian mosquitoes, or even on the classification of mosquitoes, in general. Unfortunately there was nothing to guide him as well on how to persuade laboratory-cultured mosquitoes to bite and blood-feed since the mosquitoes behaved absolutely obstinately. Finally he devised his own method whereby he soaked the bed mattress and mosquito net with water which lowered the temperature a bit and poured in some moisture as well. The trick worked for him. Dr Ross enrolled a sorely feverish patient and made him sleep inside the soaked bed with soaked bed net. Before doing so he finger-pricked the patient and noticed under microscope virtually a cascade of crescents. He got the right subject, he thought, and then released inside bed a counted number of mosquitoes who ultimately relished on the patient's blood. He collected them with the help of a test tube and bottled them for dissection almost instantaneously. Result was as ever before - disappointment.

'tis not in mortals to command success,
'll do more, deserve it.

Dr Ross had a 'never-say-never' attitude and was a man of positive thinking to the hilt. He had 'no time for repentance' on failures. He was always charged and bubbling with reinvigorated activities of one or another type; he had no tranquil life outside work. By this time he was approaching his 40th birthday and through sheer dint of hard work and focused application had developed a real scientific character since, he learnt much earlier in life, it is the Character which is the highest embodiment of the human being – the noblest heraldry of man; it is the Character which dignifies him, which elevates him in the scale of manhood, which forms the conscience of society, and creates and forms its best motive power. He was greatly inspired by Dr Patrick Manson, with whom he was in regular correspondence whereby he was reporting to him about all the daily experiments with malaria parasites in human blood and, though inconclusive, in mosquitoes.

Here it is opportune to mention that he always referred to blood-engorged mosquito as 'he' or 'him'. Some authors have painted this deficiency in Ross equating it with the fact that in armed forces during those days only men served the army and the men in uniform were habituated to refer each other by, naturally so, masculine verb. This, however, cannot be the case in case of, at least, Dr Ross for he was a highly educated person and, unless completely ignorant of the subject, could never have made a blunder like this 'in writing' about the gender difference. The matter-of-fact is that Dr Ross, by his own acknowledgement, knew almost nothing about taxonomy and biology of the mosquitoes, and hence the mistake!

"When you are inspired by some great purpose, some extraordinary project, all your thoughts break their bounds. Your mind transcends limitations, your consciousness expands in every direction, and you find yourself in a new, great, and wonderful world. Dormant forces, faculties, and talents come alive, and you discover yourself to be a greater person by far than you ever dreamt yourself to be."

— (Maharishi Patanchali)

Next few days he spent in studying mosquitoes, by dissecting stomach, intestine, malpighian tubules, and ovaries, most carefully and repeatedly continued to see spherules and sphericules only. He was obliged to conclude that the crescents of the malariated human blood transformed to some other forms and size, but was not dead sure to announce yet.

He wrote to Dr Manson, who in turn appreciated his achievements and encouraged him to continue his experiments with reinvigorated zest and accuracy. Manson wrote to Ross to follow, what the latter termed, as Manson's theory of "Grand Induction." Ross followed his advise letter-by-letter and examined more mosquitoes but saw nothing which he could swear were parasites, save for his new thesis that was taking shape gradually and irrefutably: the crescents from human blood transformed to spherical bodies in the mosquito body. Ross followed instructions and surmised that the gregarine parasites actually represented some mosquito-stage of the malaria parasite. He, therefore, decided to work out the whole matter anew systematically and experimentally down to the fullest detail. He conducted ensuing experiments with well-founded alacrity and was almost ready to verify that the free living flagella in the mosquito stomach were nothing but spermatozoa when he was served with office orders to proceed forthwith to Bangalore. This was possibly first time in Ross's military career when he became angry with the system within services. He was keeping his superiors abreast with the progress of his malaria experiments on day-to-day basis and just when he thought he had achieved a big goal he was shown the way! By twist of destiny or by his own sudden exhibition of volcanic exacerbation to the orders he, all alone in his long crusade against malaria, became, in Ross's own words, 'a militant apostle.' This, in Ross's opinion, was the first such serious interruption in his malaria research just when rapid advances were being assured. Yet, he joined on his posting at Bangalore without a murmur.

तू न थकेगा कभी, तू न रुकेगा कभी,
तू न मुड़ेगा कभी, कर शपथ, कर शपथ, कर शपथ,
अग्निपथ अग्निपथ अग्निपथ।

(You shall never tire,
You shall never slow down,
You shall never turn back,
This oath you will take today!
This oath you will fulfill in your life!
Take this oath!
And walk the Path of Fire, every single day!
The oath of fire! Yes, That Path of Fire!)

– (Dr Harivanshrai Bachchan)

● ● ●

8

Bangalore: Great is Sanitation – *"The Greatest Work, Except Discovery, that a Man can do"* (1895-1896)

"Sanitation
is more important than political independence"
— **Mahatma Gandhi,** *Father of Nation*

Way back in mid-1890s Bangalore was not a notably malarious place, but, as Dr Ross had come to know of the city during his earlier stint, it reeled continuously under the effect of cholera and dysentery in particular. He, however, turned tables in his favour in his grand style. Bangalore gave him a very unique experience in practical sanitation which was invariably associated with malaria eradication as well. From Manson's experiences about unraveling the enigma of filariasis transmission in China, he had understood that the road of research is often long and meandering, full of thorns, and only the determined and engrossed ones complete the journey. Even the best of opportunities and helps can not make the idle industrious, the thriftless provident, or the drunken sober. Such reforms can only be effective by means of individual action, economy, and self-denial, by better habits, rather than by great rights. In the order of nature, the collective character of an explorer will as surely find its befitting results in his life and health, as water finds its own level.

On his arrival at Bangalore, on 9th September, 1895, he discovered another outbreak of cholera in the making with several scores of deaths already reported during the month of August but he could easily sense the Municipality's indifference in performing their duties with enthusiasm and responsibility. There was a Sanitary Board, small though, and inadequately equipped staff- and tool-wise. It required a complete reorganization as well overhauling of its schemes of work. Accordingly, taking into account of a diploma of Public Health by Dr Ross, the Government of India had vested into him with all the necessary powers which he found astonishingly sufficient. He innovatively exercised strict public health and sanitary principles in tackling the cholera and dispatched six emergency reports on his works to the government, and extracted 5000 rupees from the Municipality for the necessary work. The was followed by a final report on 20th January, 1896. This was considered by his superiors as an important document and received lofty accolades, reading: "Dr Ross had discharged the duty entrusted to him in a most able and efficient manner.... and his excellent report ... will contribute largely to the welfare of the Station." "His work throughout had been remarkable for the zeal, thoroughness and tact exhibited", his Resident had commented, adding, "For his work no more capable officer could have been selected." Residents appreciation meant to Ross more than the "*cum laude*" mark, rendered by his esteemed Medical College/University.

Impressed by his managerial and organizational qualities his bosses asked him to serve temporarily as an Officiating Residency Surgeon under the Resident on 26th May, 1896. That year at least thrice cholera had broken out implicating several lives. Dr Ross brought in radical changes in the entire regimental and township area: water conservancy, water supply system through conduits, refusal-removal improved, laterines and drainage system, even taxation and Municipality Regulations & By-Laws were improved; Municipality-controlled laboratory was initiated, registration of vendors for food, milk, and water, and scores of other amendments were brought into force. Although these changes did not bring about any perceivable difference that whole year but the ongoing cholera was arrested immediately. For future prevention Dr Ross had prepared arduously a guideline, '*Memorandum on Open Sewage Drains*' for the Municipality under which the street gutters, which were generally remaining permanently choked, were regularly opened up to give way to the water. He and his associates had hardly begun to congratulate themselves on the result when, on 8th August, a terrible epidemic resurged in North Blackpully and the East General Bazar – teeming with overgrown and impoverished population. Dr Ross, an astute administrator now, made all the wells closed as well as disinfected them over and over again. He was much concerned about the poor people whom he suggested to drink water after boiling the same first. The reeking rain, however, played the spoilsport and washed the reeking filth of the streets and latrines into the shallow fouled pools or wells. Cholera outbreak during August 1896 was not skewing at any price while Dr Ross had a dreadful time. And then it happened, after nearly a month the pestilence subdued and finally disappeared, not before taking lives of 150 peoples and another 69 sickened; the actual number of both the dead and affected was probably several times higher!

In enforcing laws of public health and sanitation to thwart the impending onset of the deadly cholera in Bangalore Dr Ross had had to toil exceedingly superlatively, to often convince the Municipality Commissionerate whose indifferent attitude was well understood. "This in spite of", Dr Ross observed, "the revolutionary discovery of the causative agent of the cholera, *Vibrio cholerae*, by Dr Robert Koch of Germany, that too, from Calcutta on 7th January, 1884!" (In fact it was the Italian anatomist Filippo Pacini who first isolated *V. cholerae* as the cause of cholera in 1854, but his discovery was not widely known until Robert Koch, working independently 30 years later, publicized the knowledge and the means of fighting the disease. He then, of course, hogged all the limelight and laurels in consequence although his own work of isolation and identification was no less meritorious in any measure!!). Ironically, only after more than 130 years post-discovery, India has now taken up, almost after seven decades of independence in 1947, in mission mode the countrywide sanitation in all aspects of life – "The *Swacch Bharat Abhiyan.*" It has been exactly six years since the Indian Prime Minister Mr Narendra Modi launched the "Swacch Bharat Abhiyan," or "Clean India Mission," to honour India's independence leader, Mahatma Gandhi – the apostle of sanitation in personal and social life - on the anniversary of his birth. The aim of the nationwide cleanliness drive: to clean up the country by 2019, the year that marks the 150th anniversary of the birth of Gandhi, who wanted to make sanitation a priority for India more than a century ago. The Mission has of late begun to show good results. The current drive aims to end the wide-spread practice of open defecation, manual scavenging, build more toilets, improve waste management, among other goals. It is New Bharat or India in making!

Dr Ross clearly noticed lack of coordination among various different civil bodies meant to provide essential amenities to public. He was sored to see the level of corruption in public departments and did not know how then scientists could provide better sanitary conditions to live and subsist. Dejected he once lamentedly commented, "I should like to take the politicians who so often belittle and neglect sanitation and medical science – I should like to take them by the scruff of the neck and fling them into the midst of an Indian outbreak of cholera, and into the midst of the filth and dirt which their bad laws and feeble administration allow to continue! I am not talking of British politicians only, but of the Indian politicians, such as Municipal Commissioners as well – they are all alike. In this case they had allowed these slums to grow up during years of neglect; every house had its poison-pool, and science could do nothing" (Ross, 1923). He deeply implored the living standard of the local-folks, "Here is a city seething with filth and disease. What is the preaching high moralities, philosophies, policies, and arts to people, who dwell in these appalling slums – sometimes whole families of them crowded into one cell, mixed with cattle, vermin and ordure." He thus professed, "We shall reach the higher civilization, not by use of the politicians' shibboleths, methods of government, manners of voting, liberty, self-determination, and the rest, all of which have failed – but by scientific ordering of cities until they are fit for men of the higher civilization to dwell in. We must begin by being Cleansers." How true he was, more than 130 years ago even!

His prophesy proved true when around 1980 the same Bangalore city boomed with human population (>10 million), on one hand, and swarming mosquito nuisance, on the other (Rao, 1975; *unpublished Report*). Though the major nuisance was caused by *Culex quinquefasciatus* (97%), nevertheless other mosquitoes such as for example *Cx. gelidus, Cx vishnui* group mosquitoes, *Cx (Lutzia) fuscanus, Mansonia annulifera, Aedes aegypti, Ae. albopictus, Armegeres subalbatus, Anopheles subpictus* and *An. vagus* escalated the loss of tranquility. Incidentally two of these mosquitoes, *Aedes aegypti* and *Ae. albopictus*, unleased a spate of outbreaks of dengue and chikungunaya fevers in Karnataka State, including Bangalore (now Bangaluru). The problem of nuisance due to mosquitoes became so grave in Bangalore that Mr Ramakrishna Hegde, Chief Minister of Karnataka State had got to call a high level meeting on 23rd August, 1983, with the ICMR-Vector Control Research Centre to study the problem of mosquito nuisance and prepare a master plan for control of mosquitoes in the Bangalore City. A two-year long strenuous study in 151.16 sq. km Bangalore City Corporation area plus an additional 496.64 sq. km area under the Urban Malaria Scheme was detailed out in the document "Bangalore Mosquito Control: Master Plan" (Rajagopalan et al., 1987), the first ever such a guidelines book in the history of Indian Metropolices (Fig. 8.1a,b). This document was a corroboration of Dr Ross's well-judged dogma of public health responsibility, commitment and execution with absolute resolution: "A genuine campaign does not consist merely in the formation of expert committees, the passing of ordinances ... and the issue of wide advice to the public... it must always be permanent concern of the state, requiring careful measurement of the amount of the sickness present, a nice appreciation of measures most suitable for the locality, exact estimates of their cost compared with the cost of the disease, a well-considered organization, and, above all, a fixed determination to succeed."

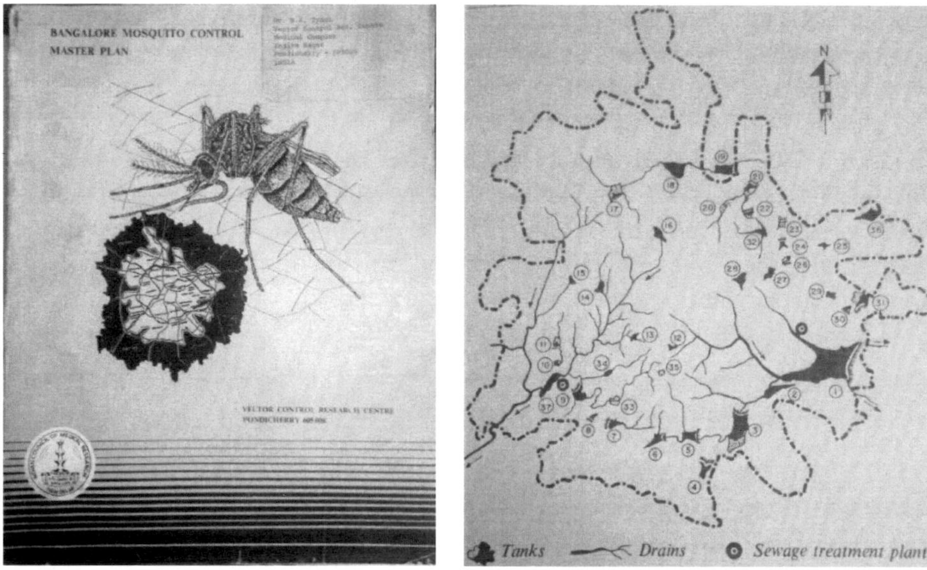

(a) (b)

Fig. 8.1a,b: (a) Facsimile of the cover page of the document, "Bangalore Mosquito Control: Master Plan", and (b) graphic map showing details of the major mosquito breeding sites; (1) Belandur, (2) Agaram, (3) Madivala, (4) Hulimavu, (5) Duraiswamypalya, (6) Sarakki, (7) Uttrahalli, (8) Arahalli, (9) Hosekerahalli, (10) Nayandaha, (11) Lorrypalya, (12) Lalbagh, (13) Kempabhudi, (14) Dasarahalli, (15) Kamakshipalya, (16) Sankey, (17) Mattikere, (18) Hebbal, (19) Nagavara, (20) Kalagondanahalli, (21) Hennur, (22) Kacherakanahalli, (23) Banasvadi, (24) Naganapalya, (25) Dooravannagar, (26) Byappanihalli, (27) Kodarampalya, (28) Ulsoor, (29) Byrasandra, (30) Annasandrapalya, (31) Doddanekundi, (32) Lingarajapura, (33) Ittamadu, (34) Pappana, (35) Yediyur, (36) Krishnarajapura, (37) Kengeri.

During his sojourn in Bangalore although Dr Ross did not work on malaria, he could not, but his enormous scientific and administrative interventions in the surging cholera-related issues he apparently imbibed unparalleled lessons in practical public health and sanitation that no book or amount of reading literature could have instilled in him. On his part, however, he continued making correspondence with Dr Manson and reported about his results on the mosquito larvae with 'gregarines' given to certain selected subjects at Begumpet on 22nd August, 1895, but everything was inconclusive and of no avail. Without any formal training he was able to distinguish two groups of mosquitoes that he was constantly working with, the brown (likely *Culex*) and the brindled (possibly *Stegomyia* or *Aedes*), both of which, as he had learnt later that each species of parasites might require a different species of mosquito – which was not strictly correct, did not carry the human malaria parasites. He wished he could be posted to work in a place where malaria parasites-bearing mosquitoes were in abundance, and he wrote about this to Dr Manson with the intention to help. He also acquainted him with his work on various avifauna, without any success. He planned to traverse across the country from Simla to the plains in the search of right species entirely on his own expense. On the hindsight, Dr Ross felt absolutely engrossed in the cholera and sanitation work and he wished he could continue to 'superintend the Bangalore sanitation for years.' His hard work finally brought fame to him; he was demanded to look after similar scenarios in Tirah (Afridi) and in Patiala in Northern India which were straightforwardly refused by the government.

●●●

9

Sigur: The Third Category of Mosquito ('C' Type) 'Dapple Wings' Sighted – *An 'Anopheles' for sure*

"Science is discoveries, inventions and innovations – which only make our understanding about life and environment better."

In March 1897 Dr Ross, a medical officer in the 19th Military Infantry Regiment, applied for a short leave for a month and joined his family in Ootacamund (8000 ft MSL), placed on the rounded tops of the Nilgiri Hills, which gave the family a feeling of a bit of England. The leave was only for namesake since Ross had tied up his vest to re-commence his work on malaria. The place lacked malaria which prevailed at the foothills in the glens and vales not far away. Dr Ross prepared for a final assault on his enemy, the deadly mosquito, though he still did not know its exact nature or skills. He explored the area and selected Tippoogadu and Kalhutti in Sigur Ghat– his venue of ambush. He was also waiting for the clear orders to work on malaria from his regiment after his leave period was over. In fact he was meanwhile expecting a lucrative posting. But, as it happened on many other occasions in past, after much upheavals of fate (he lost an appointment to the officiating civil surgeoncy at Akola, Bombay Presidency). As the saying goes, misfortunes do not come singly. He endured more than he deserved.

During his stay at Sigur Ghat he adventured several marches to the trenches at the plains, looking for malaria. Once he was seized with a severe go of fever forcing him to discontinue his march. Throughout that night he had fever with pain in the liver, followed by rigors and aching in bones and gradual defervescence. He thought it could be malaria because occasionally he slept without a mosquito net. Earlier he had noted that at one time one of his colleagues had fallen sick due to malaria simply because he had slept without a mosquito net. He was however much relieved after consuming quinine for five consecutive days. As soon as he was healthy enough again after a few days' recuperation he commenced screening fever cases for malaria parasites. He once came to see a puddle full of larvae which he collected in a bottle for examination. His mind was occupied in guessing the place where he had caught the fever, possibly a place called Westbury, Kindersley's Estate at the foot of the ghat. He walked up and down to arrive Kalhutti and, all along, his mathematical mind was busy calculating (though erroneously) estimate of parasites he might have hypothetically breathed in during his nine hours of work, viz,. 25,000,000. He carefully examined water puddles in Westbury again but did not come across a single positive site. He jogged his memory all around to find where after all he had contracted the fever and it occurred to him that he had also visited Sigur Ghat on way to Kalhutti (Fig. 8.1a,b). Back home he had a relapse of malaria bout and found the parasite in his blood.

Fig. 9.1a,b: above (a) Sigur Ghat falls, below (b) Kalhatti Falls range (*Source:* Wikipedia; Free Wikipedia)

He repeated exploring areas in Kalhutti and Sigur areas, searching malaria cases and the mosquitoes there. He never found any larvae but came across an individual adult in the rest house, engorged on blood. Carefully approaching the mosquito he caught it alive. By then Ross was quite knowledgeable about the brown (*Culex*) and brindled (*Stegomyia*) mosquitoes and quickly discovered that his catch was unique since it was clearly distinguishable from the aforesaid two types; this mosquito being small with wings striped 'brown' and white (...later he corrected the wings were 'spotted' not striped. It was the first *Anopheles* mosquito he had ever observed, but without actually knowing it by that name. He continued to call it 'grey' Dapple Wings mosquito.), His explorations into the Kalhutti-Sigur jungles continued unabated. He regularly encountered malaria cases with splenomegaly but never a larva albeit adults aplenty. All along this wrecking drudgery he could still complete a manuscript pertaining to his recent malariological studies including also the blood-feeding behaviour of mosquitoes for the South Indian Branch of the British Medical Association; many months later this was republished in *The Indian Medical Gazette* (1898). Thus, Dr Ross formed the opinion that mosquito's contact (blood-feeding) with man is obligatory. This was a new thought altogether and he immediately wrote *currente calamo* to Dr Manson. Behind his delight there was unseen disappointment too; he was worried for the continued biased considerations of his superiors (no promotions, no escalation in salaries, no freedom to work on malaria etc. etc.).

In the midst of all this physical and mental turmoil he was also appearing indecisive about his career and rather seriously giving a thought to go on retirement with first pension due in a year. But before doing so he wanted to accomplish his task of solving the Great Problem. He was much discouraged and far from being well. At that delicate moment, as always before, he looked to Dr Manson. He beseeched Dr Manson to use his influence to get him posted to an year's special duty for completing his studies as he thought he was very close to the Ariadne's thread in the humanity's efforts to control malaria. Dr Manson made it for him. He returned to Secunderabad - the cradle of the century's greatest discovery (of mosquito-malaria connection by Dr Ross) in next two months which also subsequently fetched him the world's greatest scientific honour, the Nobel Prize, in 1902!

"Though losses and crosses
Be lessons right severe,
There's wit there, you'll get there,
You'll find no other where.""

— (Burns)

• • •

10

Secunderabad: Dr Ross's Epochal Discovery: *The Ice was cut but the road ahead was like a Double-Edged Razor* (1897)

যদিতোরডাকশুনেকেউনাআসেতবেএকলাচলোরে॥

তবেএকলাচলো, একলাচলো, একলাচলো, একলাচলোরে॥

যদিকেউকথানাকয়, ওরেওঅভাগা,কেউকথানাকয়
যদিসবাইথাকেমুখফিরায়েসবাইকরেভয়–
তবেপরানখুলে
ওতুইমুখফুটেতোরমনেরকথাএকলাবলোরে॥

যদিসবাইফিরেযায়, ওরেওরেওঅভাগা, যদিসবাইফিরেযায়..
যদিগহনপথেযাবারকালেকেউফিরেনাচায়–
তবেপথেরকাঁটা
ওতুইরক্তমাখাচরণতলেএকলাদলোরে॥

যদিআলোনাধরে, ওরেওরেওঅভাগা,
যদিঝড়–বাদলেআঁধাররাতেদুয়ারদেয়ঘরে
তবেবজ্রানলে
আপনবুকেরপাঁজরজ্বালিয়েনিয়েএকলাজ্বলোরে॥

যদিতোরডাকশুনেকেউনাআসেতবেএকলাচলোরে॥

If there is no-one responding to your call - then go on all alone
If no-one speaks (to you), don't think you are unfortunate,
If everyone turns away, if everyone fears (to speak), then with an open heart without hesitation speak your mind alone
If everyone walks away, O unlucky one, everyone walks away
If no-one looks back towards the (your) unpredictable path, then with thorn pricked (of the path) bloodied feet, walk alone
If no-one heeds your call - then walk alone
If no-one shines a light (on the path), O unlucky one.
If the dark night brings a storm at the door - then let the lightning ignite the light in you alone to shine on the path
If no-one heeds your call - then walk alone and let it burn alone."

–(**Rabindranath Tagore**, Nobel Laureate)

The famous maxim "Courage is the thing, all is gone when this is gone" applies letter-by-letter to Dr Ross for he, despite gravely unfavourable situations, was always ready to push forward and brace new challenges. His courage was more truly heroic than the achievements of physical valour which are rewarded by honours, titles and laurels. His kind of courage displayed itself in silent effort and endeavour, enduring all and suffer all for truth and duty. The character embodying the steel of courage was steeped in blood of his past lineage - full of energy. Energy is the very central power of character in a man – in a word, it is the man himself the scientist way. Energy enables a scientist to force his way through irksome drudgery and dry details, and carries him onward and upward in every station of life. It gives impulse to his every action and soul to every effort. True hope is based on it, - and it is hope that gives the real perfume to life. It accomplices more than genius, with not one-half the disappointment and peril. There is indeed no blessing equal to the possession of a stout heart. So, when a scientist fails in his efforts, it gives a satisfaction to him to enjoy the consciousness of having done his best rather than lament the negative result. Failures, in the first place, are a teacher to a scientist who thus develops a combative skill against all odds in his humble yet determined pursuit of research, moving forward inch-by-inch anchoring to patience, and finally triumphing in his integrity. A true scientist is he who keeps walking undeterred on his courage 'even when his feet are bleeding and his limbs failing him'. What better example of combating spirit can be put forward than that of the life of Dr Ronald Ross.

He joined his regiment the 19th Madras Infantry at Begumpett (now Begampet) Lines, situated near the outflow of the Husein Saugar (now Hussain Sagar) Lake on 18th June, 1897, leaving his wife and children at Ootacamund, where, to his shattering dismay, the latter contracted contorting whooping cough after a while (Fig. 10a-c). At Hyderabad it was exceptionally dry and he whiled away in playing cricket and bicycling, and, otherwise, inditing long letters to his mentor and good guide Dr Manson: "*Company of wise men always offers good knowledge*." For next three weeks or so nothing very notable happened in his routine except that, on one particular occasion when he was challenged by one of his colleagues to give evidence for presence of malaria parasites in human blood, he miserably failed to demonstrate prevalence of the malaria parasite in the known malaria patient's blood under the microscope. He was crest fallen and ridiculed in silence. The humiliation was deep in heart that he refrained completely to mention about it in his letters to Dr Manson. Next few days he spent the time doing almost nothing in malariology and he found solace in writing poetry. Nonetheless. the continuous feeling of being drifted away from his objective forced him to find his track to tread, lest the Italians, French and Americans might take away all the hard-earned discovery(!). "One more effort and the thing will be done", he pondered pensively. The great story was ready to be unfolded.

Dr Manson had continuously advised Dr Ross to "follow the flagellum" in order to disclose the hidden truth of the future of the body. Dr Ross religiously ensued the thesis and demonstrated in 1895 experiments that the sexual bodies found in the human blood had unsuspectingly transformed into sphericals – distinctly different form, shape and size - and in many the flagilla were formed. Beyond that point he could not build a correct pathway of the flagellum in the stomach of the mosquito. Ironically, this latter part of Dr Manson's thesis that the parasite somehow get away into the ground water which, on drinking, make way into the human system was never demonstrated in any of Dr Ross's experiments since 1895. So, he thought to review afresh the whole issue of parasite biology in mosquito. He also understood that to prove his viewpoint he needed to have the 'third' type of mosquito which he had earlier encountered in Sigur Ghat – the 'dappled wing'. A major shortcoming in his seeking mosquito-related wisdom was that in spite of all efforts he could neither get to study any book or literature nor was there any person he knew he could contact for solving doubts. He had had to trust his own confidence, and by the mid-July, precisely on 16th July when monsoon had commenced, he decided to go forward, with the assistance of three natives.

Before he could have encountered the 'dappled wings' mosquitoes once again, he however continued working with the mosquitoes in his hot office-cum-laboratory; he did not use 'punkha' fearing his plesiotypes (dissected specimens of mosquitoes) might be blown away! The result was invariably failure after failure and he was even hallucinated due to overwork, leaving for home almost exasperated and continuously pondering over the issue if he were using the right mosquito group for malaria result? His microscope, too, was showing signs of 'fatigue' since its screws had got rust and the eye-piece cracked. But Dr Ross refused to stop and pushed forward, with "*Work within heart, God over head*" philosophy.

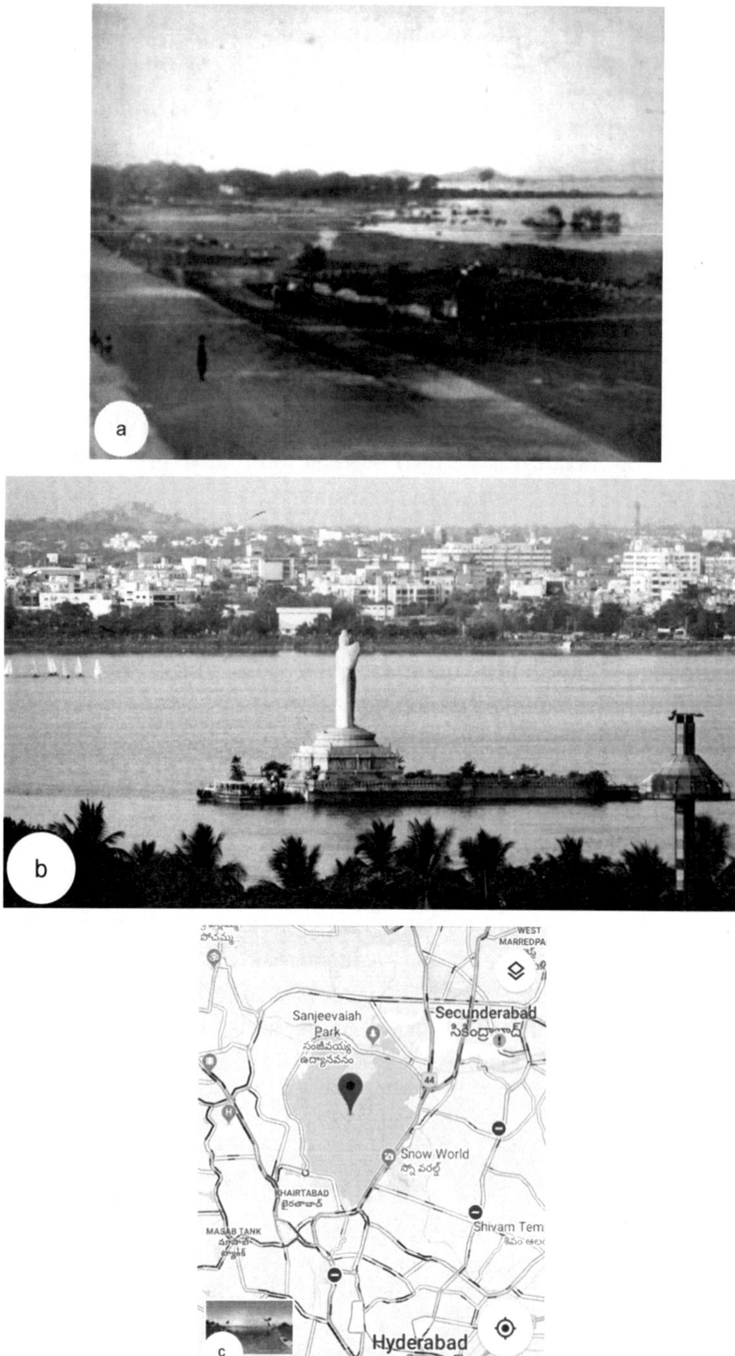

Fig. 10.1 a-c: Hussain Sagar Lake, Hyderabad in (a) late 19th century (*Photographer*: Lala Deen Dayal, in 1890, (b) same in early 21st century, and (c) Graphic presentation of Secunderabad near Hussain Sagar Lake.

As if providentially designed, on 15th August, 1897, one of his three simple native attendants brought to him a bottle containing water full of wriggling mosquito larvae some of which to Dr Ross were quite different from those of other two types, *i.e.*, grey and brindled types: *God helps those who help themselves.* Again, on 16th August, his hospital assistant, an Indian native, showed him a 'dappled winged mosquito' resting on the wall. He bottled it carefully and, after quickly incapacitating with tobacco smoke, dissected the mosquito hoping to see parasites under the microscope. Alas! He could see nothing of that sort. Failure, once again. The hospital assistant looked at his face which painted a dark shade of dismay and dejection. He almost yelled with pleasure, " Sir, I know a place where many such critters can be seen. They were the same as those emerged in the bottle"

"Where there's a will there's a way."

Leaving everything as it was in the office he dashed to the place he was guided to by his hospital assistant. He was amazed with elation to see about a dozen 'dappled winged mosquito', his 'C' type, trying hopelessly to pass through a gauze. He collected them; his first face-off with the 'dappled winged' mosquitoes ever in Secunderabad.

Hussein Khan, his hospital patient with copious crescents in blood was immediately brought to a secluded room, and made to lie down on a cot under the mosquito net. Before putting Husein on cot he was body-bandaged to restrict mobility. At 12.20 p.m. he squeezed from the inside the bed net 10 fully blood-fed, and put one each in a separate test tube. Between 12.40 and 12.50 pm, first he subjected two of his catch to systematic operation or surgery to see under the microscope any crescent-sphere flagella metamorphosis in them. He succeeded little and for a disappointingly ephemeral time. His disappointment doubled up next day morning when he found two of his gorged mosquitoes had perished. For a moment he got panicked and hurriedly dissected two of the remaining mosquitoes, but in vain! In the raze of excitement he dissected, or rather wasted, all the remainders sans any sign of the crescent-sphere flagella metamorphosis in them. The man was so obsessed with dissection that he left no stone unturned in the process and even began dissecting fleas caught from a rat for trypanosomes. He was emerging infinitely fearless and his mind was growing stronger.

Where the mind is without fear and the head is held high;
Where knowledge is free;
Where the world has not been broken up into fragments by
narrow domestic walls;
Where words come out from the depth of truth;
Where tireless striving stretches its arms towards perfection;

Where the clear stream of reason has not lost its way into
the dreary desert sand of dead habit;
Where the mind is led forward by thee into ever-widening
thought and action ....
Into that heaven of freedom, my Father, let my country awake.
– (Rabindranath Tagore)

Collecting all his energies he sent for his hospital assistant and the three native incumbents and asked them to take him to the place where one of the latter had originally found the 'dappled wing' mosquito. To his grave misfortune and sheer disappointment they miserably failed to spot the site again! Somewhat angry with the gross incompetency of his assistants, Dr Ross returned to his office-cum-laboratory and sank on his easy chair, deeply pensive. Next moment he jumped out of his comfort zone and began to jolt down details of his work to Dr Manson.

In the meantime, he found on the morning of 19th August, 1897 that the mosquito larvae fetched to him by one of his native attendants had begun to hatch in the original bottle. He fed them on Hussein Khan again and carefully selected only two of them for dissection to see crescent-sphere flagella metamorphosis in them. Failure, once more. He was left looking blankly on the wall. He became stressed for, by his estimate, time was running out of his hands. He desperately wanted results. Yet, for him, demonstrable results following repeated experiments was absolutely dogmatic. From his already shrinking reserve, he picked up a third mosquito on the third day of blood-feeding (so that the blood in the stomach was by that time completely digested), and dissected with utmost care. His dissection was not reasonably neat, yet he found there some "peculiar vacuolated cells in stomach about 10 μmm in diameter." It was certainly a "eureka" moment for him but he, like a good scientist, exercised complete restraint before doubly confirming the presence of the peculiar bodies. Dr Ross was still subscribing to Dr Manson's thesis that mosquito feed only once in their lifetime, and then die after laying eggs on water. He hoped for better results with next dissections on the remaining samples – only three mosquitoes left from that batch.

That night he tried to sleep but could not. Since he felt restless he walked out of his bed and entered the long verandah of his bungalow and quietly began to stroll up and down without making any sound of his stepping. The wind was cool and he was feeling better there in the verandah. A light drizzle occurred making the asbestos-sheeted roof drumming in harmony. He peeped out of the building up toward the cloudy sky, trying to watch his companions since childhood – the stars, but they were all seemingly disappeared behind a thick mat of clouds. At some distant tree, perhaps, a cuckoo (*Cuculus micropterus*) resonated a melody, and, on the parapet of the verandah's wall, a myriad of black garden ants (*Camponotu* ssp.) trailed in strict discipline helping each other to carry on their loads from one place to another. The space in front of verandah was all open and the ground covered with a verdant layer of grass. He charmingly noticed many a white ant or termites (*Odontoterme* ssp.) taking to wings out from their subterranean habitats, many of whom immediately locked up in conjugation for initiating the process of birth of a new generation, much to the delight of their numerous predators lurking in ambush such as dragonflies (*Pantalafla vescens, Gynacantha hyalina*), many bird sand even bats. It was then well past midnight and wanted to have a quick sleep before taking on the beast in the mosquito early morning. He was for a moment lost in the wonderment the Nature exhibits in everything organic or inorganic. He returned to his room to capture sleep but his mind continued to struggle to solve what he had just observed outside. Suddenly a thought prevailed on him that pacified all his towering inferno of thoughts and gave him a solution for his eternal woes. Then, he slowly slipped into his sleep.

कर्मण्येवाधिकारस्ते मा फलेषु कदाचन।
मा कर्मफलहेतुर्भूर्मा ते सङ्गोऽस्त्वकर्मणि।।

(You are entitled to do your duty and action,
but never to the results of your actions.
Let not the results be your motivation,
and do not be attached to laziness and inaction.)

–(Srimadbhagwadgita)

On 20th August Dr Ross customarily attended to his routine office duties; he made rounds to the patients' wards, attended to his correspondence and, of course, enquired about the mosquitoes caged the previous day for his experiment. But he got infuriated when informed that all the three ready-for-dissection mosquitoes had died last night. More so also because his native Indian attendants could not succeed in finding out a place to collect more larvae of 'dapple wings' mosquitoes. Not a man of clay but of steel he determined to see the parasites in the cadaver, but failed as expected. He then took the seventh mosquito of his 16th August blood-fed ones (*i.e.*, on 4th day after blood feeding), with just one last left behind for the test! Already, due to bad sleep last night, his eyes had fatigued and he yawned continuously. Nevertheless, his dissection, made with a phlebotomist's precision, was perfect this time; on cursory examination he did not find anything unusual and felt betrayed by his belief that the 'dappled winged mosquitoes' were the key channel to regulate parasite biology before re-entering a fresh human being. The lunchtime had passed but he continued empty stomach. Then just as he was about to rise in distress from his dissecting table he, in a rather blitzkrieg motion, opened up wide the body wall covering the stomach. "Eureka", he cried spontaneously, at the top of his pitch, since the almighty God had finally showered on him His choicest blessings. He wrote and drew in his diary what he had just gone through – the moment of triumph (Fig. 10.2 and 10.3)!.

Fig. 10.2. Facsimile of page in laboratory notebook No.1: The original record of the discovery of the malaria parasite in stomach tissues of an Anopheles mosquito, 20th and 21st August, 1897. (*Source:* PPT slides of Col. A. Kumar and SRRIP archive)

Of the "Oocyst", totally twelve in count and each measuring about 12 μmm in diameter, embedded in the stomach wall, and scrupulously identifiable due to the black pigment exactly similar to the ones in the parasite crescents in the blood. He questioned himself as to how could parasite flagella move to the stomach wall retaining the black pigment (haemoglobin) numbering about 12 to 16? He was as sure as death about those cysts on stomach wall representing the parasite, but there was a remotely sounding miniscule doubt, too. To answer this irrefutably he needed to demonstrate growth in these cysts, with the passing time, in his last specimen – the eighth mosquito. Still, as far as his concept of malaria transmission was concerned, his 'Great Problem' had been thus solved, i.e., the mosquito-malaria connection was established.

"Success is only a journey, not a destination"

That day for the first time in his long and meandering peregrination to explore mosquito-malaria interdependence Dr Ross took his dinner sumptuously and, looking at the sky, admired the twinkling stars and their numerous constillations from the verandah of his bungalow and relished the melody of cuckoo's song. The clouds in the sky had scattered and the swarms of white ants had disappeared from his view. However, the black garden ants were still working for shifting their belongings to a safer place. He felt as if the nature wanted to convey a message to him from all these viewings: There is always a brighter day at the end of a pitched dark night and, whatever misfortunes seemed to block your efforts, keep working hard with the singleness of aim. Perseverance with focused application never goes waste. After all, he thought, "Fortune favours the brave." That night of 20th August he had slept solidly in many years. The 8th and last survivor of the mosquito's batch fed on the 16th August was sacrificed on 21st August, on 5th day after feeding, and the presence of more pronounced 'cells' (oocysts; 12-14 μmm; the largest being 20 μmm) numbering twenty one in all each of which bearing the characteristic black pigments. The 'cells' (oocysts) were therefore parasites, and as they contained the characteristic black pigment, they were almost certainly the malaria parasites growing in the mosquito tissue – the stomach wall. The coordination of his mind's eyes with hands scripted an immortal verse:

This day designing God
Hath put into my hand
A wondrous thing. And God
Be praised. At his command,

I have found thy secret deeds
Oh million-murdering Death.

I know that this little thing
A million men will save –
Oh death where is thy sting ?
Thy Victory, oh grave?

Dr Ross prepared a manuscript on his discovery, titled, "On some peculiar pigmented cells found in two mosquitoes fed on malaria blood" which was published in *The British Medical Journal* on 18th December, 1897. Later he confirmed the finding in another 'dappled Winged Mosquito' (*Anopheles culicifacies*?).

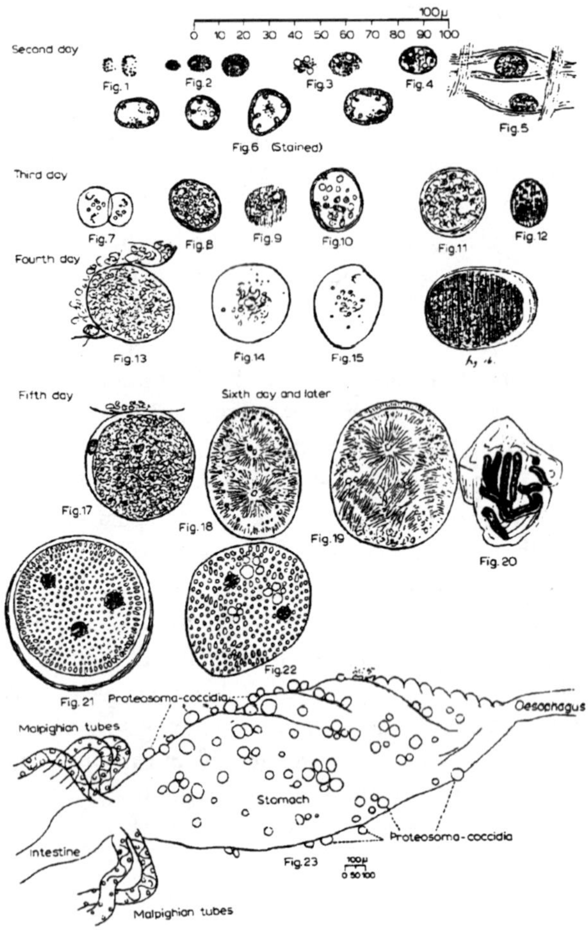

Fig. 10.3 (1-23). (adopted): Drawings of zygotes from the second to the twelfth day; (Figs. 1-5) zygotes of the second day, (Fig. 6) stained, (Figs. 7-11) zygotes of the third day, (Fig. 12) stained, (Figs. 13-15) zygotes of the fourth day, (Fig. 16) stained, (Fig. 17) zygote of the fifth day, (Figs. 18-22) zygotes of the sixth day and later, and (Fig. 23) drawing in outline of the stomach of a mosquito studded with zygotes (oocysts) of the sixth day, seen by a low power.

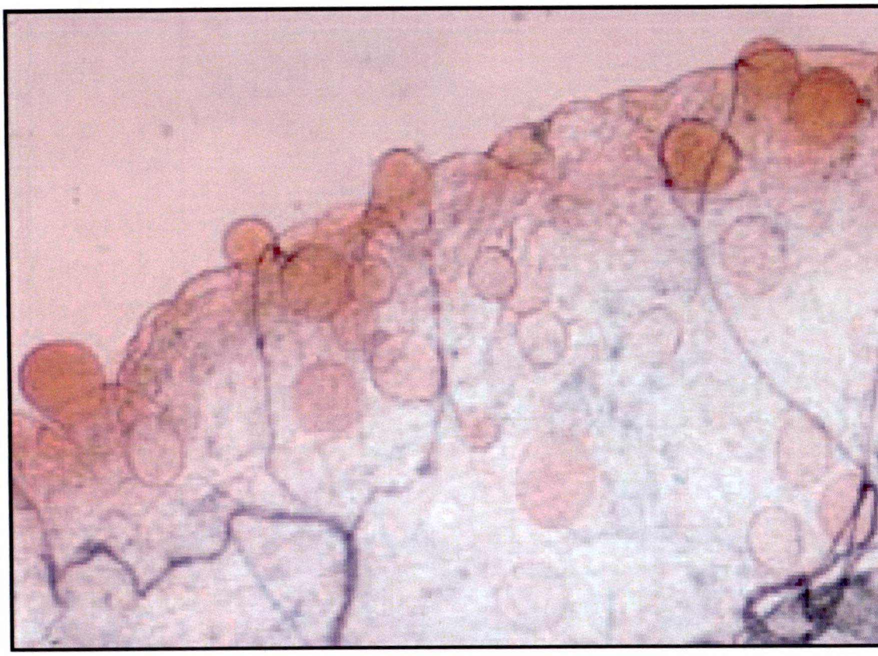

Fig. 10.4. Oocysts' stages of malaria parasites developing in the walls of a mosquito midgut (stomach). The midgut is in the lower half of the image, and the oocysts appear as multiple spheres embedded in the midgut's wall. Such oocysts were seen, for the first time, by Sir Ronald Ross on 20 August 1897. (*Source:* Wikipedia, free Wikipedia)

But, just as a query to be answered more determinedly in future, he also found seemingly malariated/dark pigmented 'cells' in the 'grey' (*Culex fatigans or quinquefasciatus*) mosquito just as he had found the filarial parasite (*Filaria* or *Wuchereria bancrofti*) in his 'Dapple Wings' mosquito (*Anopheles stephensi*?) (Ross, 1923).

"Science is dynamic and
no research is ever complete."

• • •

11

Ross's Correspondence with Manson*: *Distant Learning through all the Thick and Thin*

"Verba volant, littera scripta manet."
[Words fly away, written letters remain.]

Science is documentation of observations, and scientific correspondence with contemporary researchers is a major way of both free learning and documenting observation; their analysis, criticism, improvement, and finally reporting via known scientific channels such as for example the scientific periodicals, books, presentation of work in national and international congresses, communication of observations and/or data with fellow workers, and noting day-to-day work done with observations and data on experiments carried out. In many rare cases communication between the communicators has been reportedly built at the mind's level whereby they communicated through telepathy. The fathomless wisdom thus availed is uncommon and deeply rooted. Such true wisdom is like a deep sea - quiet and pristine! All great scientists, for that matter, are essentially men of few words; precision and accuracy being their hallmark!! Both Dr Patrick Manson – the mentor, and Dr Ronald Ross – the protégé, were essentially epitomes of stoic character bejeweled with traits such as for example positive thinking, to-the-point discussion, inexorable habit of documenting all scientific observations, and speaking about results only after reconfirming the same time and again through repeated experimentations.

"Real glory
Springs from the silent conquest of ourselves,
And without that the conqueror is naught
But the first slave."

The correspondence between Dr Ronal Ross and Dr Patrick Mansion worked like a guidance manual for the former (Gibson, 1978, 1983). They

---

* Keeping in mind the scope of the book we have limited our consultation to their correspondence made after their first meeting in London to the time when Dr Ross sought pension from the I.M.S., i.e. between 1895 and 1899.

happened to know about each other, as if almost designed so by the destiny, during the second furlough (1894-95) that Ross had availed and, there in England, on a genuine advise of an acquaintance he headed to meet with who was destined to be his lifelong mentor and guide, a *Guru*. Dr Manson, on his part, was senior, famous, influential and had a great heart of a Samaritan. He had already carved out a special niche for him in the world of progressive tropical medicine, science and research. Having served for many years in southeast Asia (Taiwan and China) researching on many a disease, but especially human lymphatic filariasis where he established an inexorable association between the infection (then known as *Filaria bancrofti*) and the mosquito (*Culex pipiens*) and, alongside his filariasis research, had a fairly good idea about the mode of malaria transmission as well. He was truly an unparalleled expert in tropical medicine of his time! On the other hand, Dr Ross, a doctor in the British Indian Medical Service, was young, energetic, hardworking but completely directionless as to his future. Though initially he had no intension to become a doctor but on his father's insistence he had entered almost reluctantly in the I.M.S.; he was interested in arts, music, poetry and mathematics. This contrasting dual character within depressed him soon. He became apparently frustrated by the persistently 'bossing' environment in his Service, and was almost raw and inexperienced researcher, questing sorely an Ariadne's thread. He was received by Manson with his usual caring warmth. Manson gave full time and ears to Dr Ross to elaborate with ease all his queries about malaria. Even though Ross was hardly knowing anything about malaria, did never before see malaria parasites and, what is more challenging, did not know an iota about mosquito types leave aside identification, he never sank in spirits!

"Pitch thy behaviour low, thy projects high,
So shalt thou humble and magnanimous be,
Sink not in spirit; who aimeth at the sky
Shoots higher much than he that means a tree."

In spite of Ross's novice in regard to malariology, Manson, a great judge of indiscernible qualities of a man, was quite deeply impressed by the raw enthusiasm and energy strongly backed up by his diverse polymathic wisdom. He visualized in Ross an emerging student, *shishya,* under him and took upon himself the responsibility to help him in every possible way. As a first step, keeping in view Ross's fantasy about malaria, Manson eagerly showed him the malaria parasite under microscope in his laboratory which, for the first time, unclouded young Ross's mind and thoughts about the malarial parasite. Manson also told him about his own discovery of the connection between *Culex pipiens* and the lymphatic filariasis (caused by *Wuchereria bancrofti* transmitted by the mosquito) and also his own 'mosquito malaria theory', though quite incorrect, of 'Grand induction' whereby he believed that the malaria parasite in man was possibly passed by drinking such water which was polluted already by the malaria parasite-infected larvae! Yet, even though Ross was seemingly unconvinced completely, nevertheless the seed of the thought about the mosquito behind the causation of malarial infection was securely implanted in his mind and he seemed to be seeing finally a ray of hope, with a long way to go - a journey which would one day bring happiness in the life of all human beings in manifold ways!

"For all your days prepare
And meet them ever alike
When you are anvil, bear –
When you are the hammer, strike"

Back to India, there came innumerable moments in the career and research explorations of Dr Ronald Ross when he felt clueless about the role of mosquito in malaria dissemination, besides constant professional difficulties and work-wise frustration, and none but Dr Patrick Manson, though in England throughout, was the only one in whom he could confide and seek solace, help and guidance for his experimentation about malaria. Manson, like a dedicated and true mentor, obliged Ross with every possible assistance in both his malaria pursuits as well as military service career. The correspondence between Dr Ross and Dr Manson is rich in both scientific and social terms, and a great teacher-taught relationship. Ross was adept at sending day to day observations and drew diagrams to describe the view under the microscope to which Manson promptly sent replies and counterfeited with documents and comments that would lead Ross closer to his research, implying a very homely and affectionate relationship between the two. This depicts an ideal scenario whereby a true form of knowledge exchange between a mentor and a mentee, through postal snail mail was upheld, at an age yet about a century apart from the one to be digital. This was only possible out of sheer commitment and dedication for the cause of human welfare.

It was Manson's "Grand Induction" which Ross set out to prove. Often, therefore, Manson was frequently writing to Ross with a dictum "Follow the flagellum". Ross himself did not know anything about the intricacies of the malarial disease and felt 'badly equipped' for the task that Manson handed him over. In the year 1897 when Ross got transferred to Secunderabad, Ross wrote to Manson briefing Manson about the conditions in which Ross was almost forced to stay at Secunderabad he wanted to be in the Madras Civil. Ross confided in Manson so much that he was seen discussing his "prospects" in a letter from 20th June,1897 and even felt sorry to have troubled him so much but it was Ross's trust in Manson that led him to discuss his career in detail too (Bynum and Overy, 1998).

Letters exchanged between the two stalwarts, sets a clear example of how Ross always jumped at every opportunity he got, to live up to Manson's expectations because he had been forever indebted to Manson for shaping the man that Ross was then and the mutual love and admiration of Manson for his beloved disciple, Ronald Ross, also reflected clearly. Not having a proper biological background, there have been instances where Ross went haywire with his trajectory of research, for which he always depended on Manson like in the letter from 18th July, 1897 where Ross wrote, "We may call it the mosquito-convention theory as against the mosquito-parasitic theory, the flagellulae being supposed not to undergo any development.... I want to know what you think of the possibilities in this direction" (Bynum and Overy, 1998). Manson also reciprocated to Ross's commitment and always backed Ross in all critical situations. Manson's fondness of Ross revealed itself when he wrote to Crosthwaite of "Ross's researches, his conviction in the theory and his wish to continue this work." Manson asked Crosthwaite for help or advice on influencing the Indian Government to facilitate Ross's demands and added that he wanted England to get the glory of teaching the world how to deal with malaria problem. He also pointed out what Ross wanted and recommended Ross as the best man in India to carry on the malaria investigation. Ross had informed Manson about the progress of his works every single day and congratulated each other on their achievements. Ross had his 'Eureka' moment when he found the parasite bodies within the last of the mosquitoes that he had with him, "pigments indistinguishable in colour, shape etc. from that of the haemamoeba", "larger and intracellular", he quoted to Manson and he found this in not just one but consecutively in two mosquitoes. It was from that day that Ross began observing 20th of August as the "Mosquito Day". Manson appreciated Ross's findings and efforts and responded in a letter dated 11th August, 1897 writing "... spore phase and with the rosette phase of the coccidia is very remarkable. I am more convinced than ever that the mosquito is an alternate host and I hope you will soon be able to prove it" (Bynum and Overy, 1998).

Despite time constraints, both Manson and Ross kept regular communications with each other and discussed the problems and the solution of the malarial parasite at depth. Manson went over the preparations Ross sent him, he also suggested Ross to "watch the flagella in a crushed mosquito's stomach and follow it up to its destiny..." (Bynum and Overy, 1998). Gradually with time they developed a bond that was par professionalism and culminated into a homely healthy relationship. Ross's ego-centric and to an extent abrasive character earned him very few friends. He did not have any group of researchers around him and was what we call a 'loner' (Bhattacharya and Bhattacharya, 2010) but Ross found his solace in his friend, philosopher, colleague and guide, Manson who advised Ross and comforted him during Ross's dark times.

Ross had arrived in Calcutta on the 17th of February and to his utter disbelief Ross witnessed more "grey mosquitoes" and proceeded with the findings of the transmission cycle of the malarial parasite in birds. He was elated to find that these insects were the commonest in Calcutta (Bynum and Overy,1998). Calcutta became the centre of his extremely recognised research work, but Ross expressed his angst against the absence of a reference medical library and enough men capable and qualified to assist him although several men showed interest to work with him. In a letter dated 2nd March,1898, Ross was anxious because he wrote, "There is considerable difficulty in finding cases. I have obtained only four poor crescent cases..." to which Manson dropped a hint to try his hand at bird malaria instead of hooking on to humans stating "Don't forget that mosquitoes bite birds and that the plasmodia of birds are just as likely to be conveyed through mosquitoes as are the plasmodia of man." (Bynum and Overy,1998). Through such trial and errors, deliberations made at length, Ross learnt a great deal every-day and the respect for each other kept growing. Ross pays his homage to Manson writing, "My one wish is that you were here to share with me the pleasure which I have experienced yesterday and today in seeing your induction being verified step by step". (Bynum and Overy,1998).

A major entity of Major Ross' characteristic was that he could never accept India or her people as his own and sought no feeling of kinship with them although he was brought up in Almora, India and had spent a considerable time there. In fact, his attitude towards Indians was quite condescending as is evident from his statement that 'The native of India is really nearer a monkey than a man.' Ross called the laboratory assistants useless and was outraged with the Government calling it "mule as regards science" (Bynum and Overy,1998; cited by Bhattacharya and Bhattacharya, 2010) and also questioned the integrity of the Indian Postal System (Bynum and Overy, 1998). It was herein that Ross started working on the transmission cycle of the malarial parasite in birds like sparrow. Ross kept Manson posted regularly of the examinations on the birds, some of which he had failed to see the same pigments that he saw in Secunderabad and owed it to the weather.

In a letter from the 21st March, 1898, Ross made a conjecture about a phenomenon he witnessed that apparently deemed curious to him and who else than Manson to share his cerebration. Ross wrote "A specimen after 72 hours contained numerous hyaline cells; in 20 minutes they had all become vacuolated..." "I am endeavouring to work backwards to the earliest forms of the parasite in the mosquito." Ross had also been asked to look into Kala-azar during this time but Ross wanted to postpone it for later reflecting his true commitment and dedication to the work he was engaged in and being encouraged by Sir Patrick Manson. Ross was beginning to fall into despair for he couldn't find the "sporulation" in the mosquitoes and wanted to run it by

Manson before despatching but it was for quite some time that Manson couldn't respond back to the letters posted by Ross. Ross, being helpless and nearly frustrated still couldn't find sporulation for he stopped looking for it but reviving his spirits he also made "a beautiful discovery" as told to Manson in Ross's letter on 5th April, 1898, "...the nuclei of the cells of the malpighian cells which are often very like coccidia after they have lost their pigment...... These sometimes escape into the cavity of the stomach and elsewhere". Manson had responded to Ross's findings in a letter acknowledging the challenges that Ross faced and yet emerged victorious in his life's work and wished him more success (Bynum and Overy, 1998).

Whilst his stay in Calcutta, Ross was sent to Darjeeling district and then to Kurseong to track Kala-azar. On his return back to Calcutta, Ross sent a letter to Manson which read, "I have just ascertained beyond doubt that birds eat mosquitos" and "I am also getting dappled-wings in large numbers now in the rain-water puddles." Ross also mentioned how the "awful heat" in Calcutta killed his mosquitoes and he himself could never be able to stay in Calcutta permanently (Bynum and Overy, 1998). There were reasons enough for Ross to be aggrieved with Indian Government as he had asked for assistance and refused. However, Manson stood like a pillar of strength for Ross when he met influential people only to help Ross the grant for his work. Only a true confidant would take a great such leap to lead a pupil to success and glory.

It was evident that Ross shared a special bond with Manson. It wasn't limited to their professional glory but it manifested itself when they shared their personal life at length through the letters exchanged where they spoke of each other's problems in depth and unitedly tried to make a haste out of it. Ross's apathy and distaste for India grew increasingly and in a letter written on 28th June 1898, Ross wrote "Fact is we are tired of India & feel that my health won't stand it much longer". In the next few letters Ross only concentrated on vocalizing about his progress with the malarial parasite work. He made advancements as Manson suggested and found the germinal rods in the insects fed on bird which Manson had already suspected and Ross also found a sort of a gland which seemed to him like the salivary gland. Manson responded to Ross's letter by congratulating him on his deed and assured Ross that his future is secure exactly how a parent would tell their child. In a few days Ross had made a full solution to the malaria problem. He opined, "Nature brings the mosquito straight to the man or birds & puts it nicely into his blood, so as to give it every chance..." Manson had bestowed upon Ross the respect and appreciation that he deserved from the rest of the world, writing an article in the BMJ (18th June,1898) 'Surgeon-Major Ross's recent investigations on the 'mosquito-malaria theory', securing Ross's priority in the discovery before others (Bynum and Overy, 1998).

Ross kept gaining wisdom from Manson as days passed and Manson gladly advised Ross on matters concerning the malarial parasite, its storage and keeping it alive. Ross even asked Manson if the name "germinal rods" was right scientifically or not and then wrote a brief description on the starting of the infection in birds. Ross also felt hugely consumed with the malaria work and decided to take a little rest from it and focus a little on kala azar and human malaria and left Calcutta for Kurseong, Punkabari and Assam. Manson knew that it would be too much to handle for Ross all at the same time. Ross confided to Manson not only with the workload but also about his personal life, grudges, understandings and demands, such has been discerned from a letter where Ross wrote "These are destined for bacteriologists who I have reason to think know nothing of animal parasitism" on the context of The Pasteur Institute being finally established at Kasauli (Bynum and Overy, 1998). Ross wasn't hesitant to comment on the knowledge of other researches about animal parasitism, despite Ross not having any background on animal science. This reflects a sort of an intellectual arrogance on his part. Moreover, in his own work on mosquitoes, Ross failed to identify the mosquito species and was even unable to use the taxonomic nomenclatures for his work on mosquitoes.

After returning back to Calcutta, Ross continued to work on the avian malaria whereas Manson warned him that Bignami had started working on human malaria because Sir Patrick Manson feared that his apprentice might not get the light if a larger and better discovery was to happen. However, Ross thought it was only of secondary importance and it was during this time that Ross started having a cold feud with the Italians. Manson had been warm and calmly explained to Ross to do his "best to prove by set example on man what holds good for *Proteosoma* holds good for the human malarial parasites" (Bynum and Overy, 1998).

Ross was also ignorant to the natives in Eastern India calling them "stupid and superstitious", although in many moments of comforts with either his family or other fellow colleagues during marches he admired by heart the beauty of Indian hills and friendliness of her people. Towards the end of Ross's research in India, he deemed fit to move back to England on the reference of Dr Manson to the Liverpool Tropical School in-spite of having conditions fit for research in India. By the time Ross left Calcutta, he even wrote to Manson from the City of Corinth that the "proteosoma" grew at normal rate, concluding that Calcutta might possibly have had some insanitary elements. (Bynum and Overy, 1998)

Notwithstanding all the conditions, Ross congratulated himself on being able to complete the task imposed to him by Manson. Whenever someone makes a discovery or achieves something of such great proportions, it can be safely assumed that the achiever will be at least content if not completely happy with his work and the results it reaped for him. The same could have been safely concluded for Ross had not some startling pieces of evidence come before us which seem to suggest otherwise (Bhattacharya and Bhattacharya, 2010).

This meteoric correspondence records in great details of Ross's research in India between 1895 and 1899, which elucidated the role of mosquitoes in the transmission of malaria, work for which Dr Ross was awarded the 1902 Nobel Prize for Medicine or Physiology – the greatest scientific honour ever - but not without passing through all the world's thick and thin particularly his battles against Professor Dr Giovanni Battistuta Grassi, another contemporary genius, about sealing priority of discovery; the contest between the two extending to the international tribunal with Dr Robert Koch, the internationally famous German microbiologist appointed as the neutral arbitrator to decide the course of action, and the Award itself deferred till 1904! Ross described the mosquito-transmission theory as Manson's 'Grand Induction', but did never actually agree his facilitation in accomplishing the mosquito-malaria experiment and finally winning the race to take away the coveted Nobel Prize.

That a meaningful and focused correspondence may sometime bring about a sea-change in a rising scientist's life, Manson - Ross correspondence is a classical example which formed a turning point in the latter's scientific pursuit of his malaria research. So, with Manson's "mosquito-malaria theory" in mind, Dr Ross, having been primed by Manson, had returned to India to his I.M.S. regiment. Dr Ross's regular letters to his mentor document the frustrations and sometimes false and overstated expressions as well as the excitement of the 19th century's greatest discovery – the mosquito-malaria connection. Dr Ross who did not find enough encouragement or resources within his career used correspondence as a means to transmit all his malariological explorations, particularly in respect of mosquito-based experiments. Manson, in turn, acted as a kind of agent in London, publicizing his findings, offering advice and seeking to use his influence to secure for Dr Ross the working conditions he so desired. Of the total 173 letters, besides those 85 written from the two decades after Ross's return to Britain, exchanged between them a total of 110 were written by Dr Ross while Dr Manson had written exactly half of these, 55. Their correspondence also records the rise and fall of a relationship, as Dr Ross's preoccupation with his place in the history of malariology led to a breach between the two stalwarts. Although various themes such as priority, nationalism, and personal vanity punctuate this letter correspondence, nevertheless, the simplicity of integrity, respect and meticulous day-to-day reporting of the minutest and trifle-looking matters within malariology reveals new insights about the golden years of tropical medicine in the late nineteenth and early twentieth centuries.

• • •

12

Calcutta: Discovery of Malaria Cycle in Avifauna – *Ross's 'Never Say Never' attitude to a Commitment* (1898)

गंगा आये कहाँ से, गंगा जाये कहाँ रे।
आये कहाँ से, जाये कहाँ रे,
लहराये पानी में जैसे धूप-छाँव रे।

– (written by Gulzar, and
sung by Hemantkumar in '*Kabuliwala*')

Dr Ross was instructed to move to his new posting, presumably an active service, just when he was lost in dreams of celebrating his 'mosquito-malaria connection' discovery and had begun embarking on but one objective to solve: How and in what form do the parasite enter into human being from the mosquito? Or, what is the life cycle of malaria parasites outside human body? He wished he could get some more time to do his mosquito research. He informally approached some seniors but none helped change the decision of the government. He wrote to Dr Manson, pleading to exercise his extensive influence and have him placed at a malarious area so that he could complete the unfinished task of biology of malaria parasite in mosquito. Dr Manson, an ardent follower of science, too felt disturbed at the instance of regularly, and especially after Ross's finding of oocysts in mosquito stomach, displacing him in the disguise of government rules and/or of departmental emergency and not giving value to his phenomenal success in malaria.

In September 1897, Dr Ross was transferred to Bombay, from where he was subsequently sent to a malaria-free Kherwara in Rajputana (now Rajasthan). Kherwara (name derives from the large number of Kher or Khair trees, *Senegalia catechu*, in the region once upon a time) is a town in Udaipur district just 80 km from the Rajasthan-Gujarat border. The topography of the area consists of medium to high rocky hills, contours, plains and is surrounded by Aravalli Range from north to south. The hill have forests that are an essential source of income and help sustain the economy of the local tribal inhabitants. The area is semi-arid with an average annual rainfall of about 650 mm (26 in). Most rain falls during the monsoon season from July to September. Winter temperatures range from a low of 2 °C to a high of 25 °C. Summer temperatures range from a low of 20 °C to a high of 42 °C. Relative humidity is above 70% during the monsoon months but below 20% during the months of March through May. His hospital was small with minimal facilities, and virtually no malaria in the area. He had lot of time to while away but his heart was longing to see mosquitoes to return to his dissecting table. He began to get depressed and gave vent to his feelings in the verses "*In Exile*". He was told by his superior in that 'delectable' place, without any truth though, that he was rewarded with this transfer as a punishment and only officers in disgrace were sent there. It is worth mentioning here what Dr Ross personally felt with such matters of transfers, as he wrote, "It has been often suggested that I was the victim of someone's personal animosity, or of jealousy – as Surgeon Colonel W.G. King had been, or as Mr Haffkine was to be some years later, but I scarcely think so." This was the kind of restraint he exercised in expressing his mind and thoughts which in fact became his hallmark in his entire service matters; no amount of movement exasperated him at all but, of course, he got sometimes worried about his research on malaria. Nonetheless, frustrated of lack of work and facilities in Kherwara he threatened to resign from service as he felt that it was a death blow to his pursuit. Meanwhile, in England, Dr Manson had already taken up a stand to help Dr Ross and get him out of Kherwara, and instead got him posted to some other place where he could complete his unfinished task of whole life cycle of the human *Plasmodia* in *Anopheles* mosquito. During his short stay in Oodeypore (Udaipur), Rajputana he was of course bitten by an *Anopheles* (possibly *An. rossi=subpictus*) mosquito. It was only on the intervention of Dr Manson that the government of India arranged for his transfer to Calcutta on a "special duty". Ross considered this abundantly rich opportunity as a God-gift coming to him in recompense for the tribulations of Kherwara (Rajaputana). Without an inkling to him this posting to Calcutta was to open a glorious time, during which the amazing story of malaria would be unrolled little by little following Manson's 'Great Induction' theory.

"A good teacher is
Like a candle
That consumes itself
While lighting the way for others."

Fig. 12.1. Entrance to Presidency General Hospital (*Source:* Dutta, A, 2009).

On 17th February 1898 he arrived Calcutta (now Kolkata), to work in the Presidency General Hospital (Fig. 12.1). As seen now the Presidency General Hospital is a red brick building located near the famous landmark Victoria Memorial in Calcutta (Kolkata) and houses the National Research Institute. Built in 1707, it was the first hospital to be built in the city and initially was open only to officers of the British India army. It was not until 1770 that non-Europeans were allowed admission. Following the independence of India in 1947, the hospital was renamed in 1954 the Seth Sukhlal Karnani Memorial Hospital and, later, in 1957, the Institute of Post Graduate Medical Education and Research.

In the hospital he was assigned the dilapidated laboratory of a recently retired physiologist (Fig. 12.2). Dr. Ross engaged three assistants, namely, Kishori Mohan, Mohammed Bux and Purboona, on his own expenses. He explained duties to the three assistants but Purboona disappeared after the first payday since he felt frightened due to the type of work involved. Ross got cautious about the remaining two and trained both Kishori Mohan, an educated technician who remained throughout very loyal to Dr Ross in all his future research work on mosquitoes, birds and malaria, and Mohammed Bux who was a particularly fortuitous choice as he became quite devoted to Ross and was singularly adept at the work. He was so devoted to the duties assigned to him that he would even sleep on the floor of the laboratory at night to keep stray cats from killing the experimental animals. He climbed through the sewers, the drains, the stinking tanks that abounded in Calcutta and brought back all kinds of mosquitoes to feed on patients but all the human experiments were a failure. Ross was terribly dejected and turned his attention to the study of bird malaria. His laboratory soon became filled with the live sparrows, larks and crows that Mohammed Bux had brought. Into the cages covered with mosquito netting went the mosquitoes. The altogether new line of research was to him like an endless ocean in which he was left to sail without an oar. He felt completely novice and did not know where would the whole work lead to? The Calcutta explorations were further mystified like a stranger he was to the 'City of Joy';

River of faith flowing in the City of Joy,
where I got my first freedom and touched the divinely inspired soil—
Champagne, music and dance
left behind in the white man's town,
Filth, flies, lime traders and noisy crows
in the native's colony,
Unknown enemies flying in the dark,
their singing swarms in a stinging swoop
drunk on wine-red blood murdering millions,
seeking out a crusader alone in an alien land,
who struggled with 'tears and toiling breath'
to unravel the mystery behind the massacre,
with Calcutta as witness and muse.

– (S. Bhattacharya; in "A crusader in my city")

Not one who would accept lying down he immediately carried out research in malaria and kala azar, for which he was assigned. He was given the use of Surgeon Lieutenant General Cunningham's laboratory for his research. His laboratory was a small isolated building containing only two rooms, office and a verandah and was situated in a compound full of trees and bushes. When his family arrived in Calcutta, he lived in a boarding house at 8 Harrington Street. But he initially stayed at the United Service Club at Calcutta and then with Dr. Maynard, close to his laboratory. His working conditions in India were quite constrained due to inadequate infrastructural facilities and administrative support. He wriote, "Imagine my disgust on arriving here (Presidency General Hospital, Calcutta) to find that there is no reference medical library. I don't know what to do for some literature I want." The typical mindset of Ronald Ross has revealed itself comprehensively in letters to Sir Patrick Manson, dated March 9, 1898, Calcutta, "The laboratory assistants are useless and I am training men I have hired." It is from 173 such correspondences with Manson that the other side of Ronald Ross unfurls itself. He had no success with malarial patients because they were always immediately given medication.

Therefore, he built a bungalow housing a laboratory at Mahanad village at outskirts of Calcutta, where he stayed from time to time to collect mosquitoes in and around the village. He employed Mahomed (or Muhammed) Bux, Purboona (who deserted him after the first payday), and Kishori Mohan Bandyopadhyay as laboratory assistants (Fig. 12.2). As Calcutta was not a malarious place then, Manson suggested to him to use birds, as being used by other scientists such as Vasily Danilewsky in Russia and William George MacCallum in America. Manson kept him abreast with others' works. In a letter dated the 17th November 1897 Manson had already informed him that W. G. MacCallum in America had made a discovery regarding the motile filaments, showing independently that they are of the nature suggested by Simond's work. Ross, always beset with a feeling of insecurity of his place in science, was alerted toward his most significant future research and settled with a blitzkrieg pace.

Fig. 12.2. Dr Ronald Ross and Mrs Ross, with laboratory assistants Mahomed Bux, Kishori Mohan and Purboona at the laboratory in Calcutta where the life history of the malaria parasite in birds was fully worked out in 1898. (*Source:* Wikipedia; Free Wikipedia).

In Calcutta Ross's delight had no bounds noticing several varieties of dappled-winged mosquitoes, besides many kinds of grey and brindled mosquitoes, actually within the laboratory, and found the breeding-places of the latter just outside. The next thing was to obtain cases of malaria but owing a raging plague outbreak in the city the government's attention was entirely focused on implementation of Haffkine's prophylactic inoculation. Ross thus met by an unexpected and most unforeseen misfortune to focus his attention toward malaria due to extreme shortage of time. With poor interaction with the physicians of the Calcutta hospitals he evidently found no cooperation for using their cases for his experiments. Unfortunately he too had no hospital of his own as in Secunderabad and Bangalore, and was forced to induce patients to come to him on payment. Calcutta was then not very malarious, especially at that time of the year. Finally, several beggars with fever were induced to come to him on large payment at his personal expense but fled without a word when engaged for finger-prick in order to examine their blood. This placed him in complete perplexity and dismay as to what to do, until he remembered the malaria of birds.

A number of crows, pigeons, weaver-birds, sparrows, and larks were then immediately procured, and experiments commenced, with the help of his three assistants, particularly Kishori Mohan. Ross was aware that despite parasites of birds differing from those of man in some very small morphological details, they were however remarkably closely related to those of man. He followed Manson advice that a particular mosquito species could transmit only a specific malaria parasite, and he needed to first test this conjecture. Therefore, it was first advisable to see whether mosquitoes would not carry one or both of the avian parasites. A number of crows and pigeons had been found to contain *Halteridirium*; but without waiting to examine the other birds, he placed one crow, two pigeons, four larks and six sparrows in several cages all within the same mosquito netting, and then in the evening released within the net a number of grey and brindled mosquitoes bred from the larvae in captivity. Next morning many of the grey mosquitoes were found gorged and were collected and kept for several days according to his rules. On the 13th and 14th March he dissected them one by one. When thirteen had been examined with negative results Ross began to fear that he had committed yet to another tedious search for the proper kind of host of the avian parasites. But fortune was kinder on this occasion; the fourteenth mosquito had pigmented cells precisely similar to those which he had found in the dappled-winged mosquitoes fed on patients with crescents in his Secunderabad laboratory an year ago: "*Fortune favours the brave.*" Next he examined the larks and sparrows used in this experiment together with the crows and pigeons, and found that they contained not *Halteridium* but *Proteosoma*; so that it was doubtful from which kind of parasite the pigmented cell had been developed. Consequently he then put the birds with *Halteridium* in one net and those with *Proteosoma* in another, and released within both nets numbers of grey mosquitoes bred in the same bottle. Of thirty-four of these fed on the birds with *Halteridium* all were negative; but out of nine fed on the birds with *Proteosoma*, no less than five contained pigmented cells. The cells were in the tissues of the insect and he thought that the parasite could therefore be able to make its way into and live in mosquitoes; precisely similar cells had been found in mosquitoes fed on men with malaria. The chain of proof was complete. This result was obtained on the 20th March and practically proved the mosquito theory of malaria. As wont of him, Ross began to question self, "Is there a destination beyond this point in his research because the fact of basic fundamental importance is that the pigmented cells in the mosquitoes are indeed derived from the parasites in the birds. Did the whole matter not require the most formal and rigid proof especially since no life-history of a protozoal organism able to transfer itself from one host to another was then in knowledge?"

Ross prepared himself for a long and tedious series of tasks with the differential bird malaria experiments. He bred grey mosquitoes from the larvae in captivity, then were fed (a) on birds with *Proteosoma* and (b) on birds without *Proteosoma*, for comparing the results. Out of 245 grey mosquitoes fed on birds with *Proteosoma*, 178 (72%) contained pigmented cells, while out of 249 of them fed on blood containing other parasites or no parasites, not a single one contained them. Ross did not stop at it and performed another experiment with three sparrows; one with no parasites, one with a few *Proteosoma*, and one with many *Proteosoma*. They were placed in separate nets, and numbers of grey mosquitoes from the same breeding bottle were fed simultaneously but separately on them. Ten mosquitoes fed on each bird were then examined, and the total number of pigmented cells in all of them were counted. Results, hastily procured though, were a revelation; no pigmented cells were found in the ten mosquitoes fed on the sparrow without parasites; 292 in the ten mosquitoes fed on the sparrow with a few *Proteosoma*; and 1009 in the ten fed on the one with many *Proteosoma*. Ross wasted no time to communicate results to Manson who corrected his figure as 0, 571, and 1084 pigmented cells in the three sets of mosquitoes separately. At last he achieved the target of proving the fact, and the theory that the parasites of malaria develop in mosquitoes was finally practically established.

But still the parasites were in the body of mosquitoes. Where and how did they go finally to prove the point that mosquito really transmitted the parasites. He recalled that Manson had always thought that a few days after her meal of blood the female mosquito laid her eggs and died; at this moment he considered both filariae and malaria parasites escaped into the water from the insect. Ross had accepted this view, but had frequently observed that the insects did not die immediately after laying their eggs; and after watching the pigmented cells growing larger and larger without apparently ripening, even five days after the insect was fed, it occurred to him that the mosquitoes needed a feed once again! He, therefore, fed his infected mosquitoes a second and a third time, and more; and found that he could easily keep them alive for a month. This enabled him to work out the development of the malaria parasites completely. But the incessant perseverance also began to take toll on his health and he felt exasperated. He sought a ten days' leave to the Himalayan hill-station Darjeeling, harbouring also several intensely malarious spots at the foot of the Darjeeling mountains. He left Calcutta for Darjeeling on the 17th April. In here at Punkabari he completed his entire malaria report for the government and wrote a couple of manuscripts. On 4th June he returned to his seat at Calcutta, to resume his research work.

At Calcutta was maintained the *status quo* alluding to the availability of malaria patients for his work, and therefore he proceeded at once with the life-history of *Proteosoma* in birds. First, he realized that so far he was dissecting mosquitoes in water or a weak solution of salt, and he never stained his preparations. He, therefore, used a strong salt solution and the secret was revealed. The contents of the mature pigmented cells did not consist of clear fluid, but of a multitude of delicate thread-like bodies, which, on the rupture of the parent cell, were poured into the body cavity (haemocoel) of the insect, and which were evidently spores. To himself he asked a question, "What happened now to these spores?" Several related queries continued to reverberate in his mind, "Did they escape into the water as the great Manson thought or were they voided by the intestine as he had himself held the belief or even did they in some mysterious manner worked their way into healthy persons during blood-feeding as opined by King and Bignami? Ross had a big cliff to mount, all alone. He saw that the thread-like bodies were soon scattered by the insect's circulation all through its body. He could not understand a bit beyond this point and thought the end of his work. Come July 2, however, he found in the thorax of a mosquito a large cell which, surprising to state, contained within it several of the thread-like bodies. They were motile and wriggling around to enter into an altogether new entity. On the 4th July, while working upon another mosquito, he fortuitously found that the threadlike bodies seemed to become more and more numerous towards a point in the thorax - as if they were converging toward same destination. They were attached to a duct and were all contained within the same gland. In all these cells there were hundreds of the thread-like bodies, floating loosely at all angles to each other. He spotted similar spores another lobe of the gland closeby. Another 'Eureka' moment awaited him, but he was cautious to know the host gland first. On the 8th July the mystery was solved. The gland lay in the neck and upper thorax the throat of the mosquito, called salivary glands. These by nature secrete the salivary fluid which the mosquito injects in the wound made by her in the skin, to dilate the vessels to prevent coagulation of the blood and a host of other functions. The exact route of infection of this great disease was thus revealed. But Ross self-infectively pondered that still all this was inference only and the last proof was still demanded.

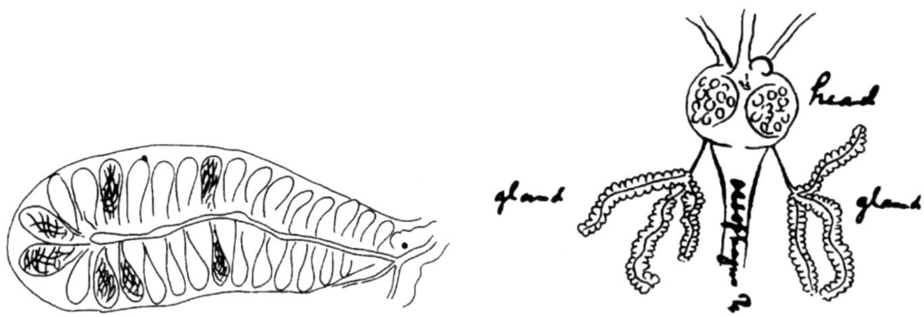

(a) **(b)**

Fig. 12.3. (a) Sporozoits in the cells of a salivary gland, (b) the salivary glands of a mosquito. (*Source:* SRRIP archive).

On the 25th June, he set out to examine again all his uninfected birds which were still quite healthy. He released a large number of grey mosquitoes, which had been long previously fed upon infected birds and many of which had been found to contain the thread-like bodies in their salivary glands, within a mosquito net in which five healthy birds were placed. On the following mornings it was satisfactorily found that the infected mosquitoes had perfectly gorged themselves freely on the birds. Out of 111 wild sparrows examined the parasites were seen only in 15 (13.5%). Infected birds, too, scarcely had parasites. On the 9th July he put to examination the previously healthy birds all of which, to his delight, swarmed with *Proteosoma.* But not content even with this he repeated the experiment over and over again and within the next few weeks succeeded in infecting 22 out of 28 healthy sparrows (79%), and also a crow and four weaver-birds, and, moreover, gave a more copious infection to four sparrows which previously contained only a few parasites. Ross, scientifically aware, kept the controls of a number of healthy birds in mosquito-nets, safe from the bites of mosquitoes, of which none (save for one exception) became ever infected. His infection experiments on birds were completed early in August, and he concluded that malaria is communicated only by the bites of mosquitoes (Dutta, 2009). At this important but much scientifically elated and elevated juncture he was ordered to move to Assam which was reeling under a quaint disease called Kala-azar (Black fever). Though Ross had completed his bird malaria research he nonetheless wanted to defer his transfer by a few more months so that he could do his human malaria related researches. The age old disease of malaria was annually taking several millions of lives in India alone and he rightly hinted as much to the Director General, but was told that he was expected to adhere to the programme. He left Calcutta on a happy note, fully satisfied for his discoveries and conclusions about mosquitoes and malaria, on the 13th August for his new duties on Kala-azar – a rather new disease which was exciting and was taking some thousand lives annually in Assam. After Calcutta's hard work, he felt much exhausted by work, heat and, of course, mosquitoes!

• • •

13

Ross versus Grassi: Malaria Controversy – *The Debate Continues*

"For all your days prepare
And meet them ever alike
When you are anvil, bear –
When you are the hammer, strike"

Tropical nations, many of which either poor or slowly developing rich, across the world have for long been affected by several deadly and/or debilitating vector-borne diseases. One of the main reasons considered behind their poverty was malaria, or paludism (in French), which is a disease that has been the scourge of populations in tropical and subtropical areas of the world, particularly India, since antiquity. The Italian name, *malaria*, identifies the disease better than the French term, *paludisme*. In fact, it refers to *mala aria*, the bad air, the miasmas evaporating from the stagnant waters of marshes, which the ancients believed were the origin of the disease. It was not until the second half of the nineteenth century that scientists started to search for the microscopic organism behind the scourge. By then, adequate optical instruments for this purpose had finally become available, and foundations for scientifically based clinical microbiology were laid by a trail of extraordinary scientists like Dr Robert Koch (1843–1910) and Louis Pasteur (1822–1895).

The reason why malaria became the focus of research was not the same for all of the individuals involved in the quest for its causal agent. France and the British Empire had expanded their colonies throughout regions of the world where malaria was the most serious and debilitating of the many parasitic tropical diseases, a factor that led to limited exploitation of natural resources and, accordingly, lower economic profits. Therefore, prominent scientists decided to tackle the problem of *paludisme* in France and in Great Britain. Malariological research also thrived in Italy, both from the clinical and the public health points of view. But Italian malarial studies had a different perspective. For French and British researchers, malaria was a "tropical disease", a "colonial" problem. For Italian physicians, it was a disease endemic to their country and a scourge that hindered the development of many southern areas of the country, which had achieved national unity only a decade earlier.

Rome, the capital city of that young kingdom, was prey to malaria in the summer and autumn months.

During mid-19th century three countries where scientists were enthusiastically devoted to malaria included, England, France and Italy, with United States of America a little while later. Among all the French and British scientists who carried out research on malaria, the most outstanding were Dr A. Laveran, Dr Patrick Manson, and Dr Ronald Ross. Charles Louis Alphonse Laveran (1845–1922) was a military physician in Algeria when, in 1880, he described malarial parasites in the blood of patients during malarial fever episodes. He called this microscopic organism *Oscillaria malariae*. The discovery was first met with scepticism. However, when Laveran gave up his military career in 1896 and resumed his studies at the Institut Pasteur, he obtained new, indisputable evidence that was later confirmed by foreign researchers. Because of his discovery, Laveran is considered the father of protistological parasitology.

In 1879, Dr Patrick Manson (1844–1922), a Scottish physician who had worked in China for over 20 years, demonstrated that so called Bancroft's filaria (*Wuchereria bancrofti*) was transmitted via the bite of a mosquito of the genus *Culex*. He proposed that a mosquito of that genus should be also involved in malaria, but as part of a bizarre cycle, the parasites were released in water by dead mosquitoes and then transferred to humans when they drank the water. Dr Manson had known Dr Ross who sought his guidance and help with his malaria research in India where he was a military doctor. He corresponded extensively with Ronald Ross, including 173 letters (compiled in a book by Bynum and Overi, 1998) in which he followed and steadily guided Ross's progress in the study of malaria in India.

In 1897-98, Dr Ronald Ross, a young surgeon in the Indian Medical Service (IMS), incriminated mosquitoes in malaria transmission, laying foundation of a remarkable piece of research considering that he did not know what he was looking for in his mosquitoes or even if they were the "right" insects. Dr Ross's discovery of mosquito-malaria connection was doubtlessly one of the greatest researches of the 19th century but it was not the solitary research on this subject. Italian scientists, in particular, had been working on malaria for many years from a time when Dr Ross did not even think of malaria as a research career. They were led by an extraordinary genius by the name of Professor Dr Giovanni Battista Grassi, already a world authority on the life history and societies of the white-ants (Termitidae), and on the developmental relationship between *Leptocephalus* and the common eel, and other muraenidae. He was a specialist animal taxonomist as well. Unlike Dr Ross who did not know anything creditable about the mosquitoes, Dr Grassi already knew taxonomically about the mosquitoes (Family Culicidae). As to mosquito-malaria research he is credited for having successfully demonstrated in 1898 that the *Anopheles* mosquito (*Anopheles claviger*) carries the *Plasmodium* parasite of malaria in its digestive tract. Because he had not cited the work on the same theme by Dr Ross, the latter took Prof. Grassi's action as a deliberate attempt to hide his pathbreaking medical discovery and a war of words razed as a towering inferno between them. Both, who were earlier good friends, became each other's deadliest foe. While it is true that Dr Ross's discovery of mosquito-malaria connection was not taken serious by for some time, nobody dared to repudiate any of the findings of Prof. Grassi – simply for the fact that he was absolutely meticulous and systematic in his experiments and conclusions; he pinpointedly determined for the first time that *Anopheles claviger* was transmitting malaria, hence the much specific "malaria-*Anopheles* mosquito connection." Ironically, the Nobel Committee had originally recommended the names of both Dr Grassi and Dr Ross, but due to the bitter and humongous opposition by the latter on the ground of principle of priority of research on the mosquito-malaria connection (Dr Ross often termed other scientists pirates on his great piece of scientific investigation – the Great Problem), and for a trail of events, scientific and political (unscientific), the decision eventually went in favour of Dr Ross. Till their last breath both these great scientists remained at loggerheads, with varied criticisms for each other's researches.

All the extraordinary research works are generally bound to attract criticism. Novel and epoch making ideas and thoughts are always subject to fierce criticism, repudiation and innuendoes. Such criticisms belong to two categories; one, that belongs to the principles and grounds on which an innovation is based is welcome, and the other one, highly treacherous with an avarice in view, that only aims to destroy the very fibre of the discovery largely influenced by the critics' own fears of being eclipsed or eliminated with their outdated and untenable dogmas! A true scientist is, however, never affected by such claims of scientific charlatans of the latter type, and keeps moving determinedly on his identified path to invent and discover new products and processes. He knows, after all, that intensity of opposition is the measure of his popularity!

Criticism in its most vicious form lurked in wait for Dr Ronald Ross who had spent several years of investigation and research soon after he discovered the association of immense medical importance. He had achieved this feat only after repeating and verifying his experiments again and again, probably anticipating the opposition he might have to encounter from the profession on making known his discovery – almost single handedly, in blitzkrieg and without sufficient resources! The route in which he at length announced his results through Dr Manson's immaculate and unassailable support in England (where he regularly kept the scientific community acquainted from time-to-time, about the advances in malaria research by Dr Ross, in India) and scientific manuscripts published in reputed journals was a most modest one, simple, perspicuous and conclusive. It was nevertheless received with lukewarm enthusiasm, as the utterances were pouring in for a rock-solid evidence for a similar theme in Grassi's approach.

Giovanni Battista Grassi was born in Rovellasca near Milan. Even though he graduated in Medicine at Pavia, he felt driven to become a Zoologist. After completing his training at the University of Heidelberg, Grassi became Professor of Zoology at Catania (1888), where he began to study malaria in birds in parallel with the Dr Patrick Manson's pathbreaking medical research on lymphatic filariasis in China and the proof of evidence for the mosquito (*Culex pipiens*) transmitting the disease. In 1895, he was appointed as Professor of Comparative Anatomy, at La Sapienza, Rome, where he met a group of malariologists. When he heard about few scientists' unspecific hypothesis alluding to mosquitoes being also involved in the transmission of malaria his colleagues in Italy argued with him and tried to convince him of the validity of *Plasmodium* transmission through a hematophagous insect - a hypothesis that he had considered doubtful until then. On the basis of malaria epidemiology and the distribution of mosquitoes in malaria zones, he focussed his investigations on three species, suspected of malarial transmission, viz., *Anopheles claviger* and two *Culex* species. Grassi's approach was a true zoological research: systematic comparative and experimental. On the 4th July, 1898, Dr Ross, while working in India, showed sporozoites, migrating to salivary glands of mosquito. He was not a zoologist and described the insect as a "dapple winged" mosquito - a term absolutely invalid for Linnaean nomenclature. In November, 1898, Grassi announced to the Lincei Academy that, along with his two colleagues, Bignami and Bastianelli, he had infected a human volunteer, by exposing him to bites of 3 mosquitoes. On 22nd December, 1898, he described the entire developmental cycle of *Plasmodium*, in the body of *Anopheles claviger* and humans, inditing absolutely scrupulously that it corresponded to what Ross had described, for *Proteosma* in *Culex pipiens*, in birds, though. However, since he did not (and quite naturally so, since he had already elaborated his aforesaid work in his manuscript) give Dr Ross due credit in his publication, Dr Grassi's action was taken by Dr Ross as an affront, inciting his fury.

Close to the end of 19th century advertisement for the institution of the Nobel Prize, commencing 1901, had already made a great message among scientists across the world and there was a fervent race among malarilogists, in particular, to increase their prospects of winning the one. Convinced that his researches were pirated, Ross began a defamatory campaign against the three Italian researchers at the end of 1900, claiming priority of discovering malaria transmission mechanism. Dr Ross began a tirade of attacks on Dr Grassi by writing to every influential person he had known, to the scientific journal and to Dr Manson, with an intention to show that his marvellous piece of research was hijacked and he alone deserved the credit. Dr Grassi answered equally viciously, discrediting Dr Ross – *hook, line and sinker*! The Swedish Academy of Science had shown an intention, to award the Nobel Prize in recognition of this discovery of enormous importance, and was therefore very embarrassed by the dispute. It appointed Dr Robert Koch, the famous microbiologist who discovered the causative agents of cholera and tuberculosis, a "neutral" arbitrator of great authority. Dr Koch nurtured a serious complaint of many years ago against Dr Grassi when he visited Italy to enquire about malaria but Dr Grassi did not pay him the kind of attention he thought he had deserved. More so, Dr Grassi had also refuted Koch's hypothesis, that larvae of *Anopheles* mosquitoes inherited malarial infection from their mother, by a series of scientific experiments. At the same time Dr Ross maintained a pleasant relation with Dr Koch. The result was that Koch's arbitration went in Ross's favour and the Nobel Prize was awarded to Ross in 1902.

Ronald Ross (1857–1932) is the main character in the history of British malarial research. Born in India, he joined the army as an officer of the Indian Medical Service. In 1897, when he studied malaria in birds, he described oocysts of the malarial parasite in the walls of the stomach of an unclassified mosquito ("a dappled winged mosquito"), which he thought was probably *Culex*. This was the starting point of Ariadne's thread that would eventually lead to the exit from the labyrinth of malaria.

A map of Italy published in 1882 indicated in red the areas with widespread malaria and in yellow the areas where the disease was present. The red area included vast coastal areas of Tuscany (Maremma), Latium (Roman plain and Pontine marshes), and Campania. Also at high risk of malaria were the Venetian lagoon areas, the Po River Delta, the Ionian Coast of Calabria, and the coasts of Sardinia and Sicily. Even more tragically amazing is another map, published in 1899, that was produced by a professor of hygiene, Augusto Celli, which indicated those railway lines where the risk of contracting malaria during a train trip was high! Italian physicians thus experienced malaria as a daily domestic tragedy, and were highly motivated to solve the mystery of the origin and transmission of the disease. The death from malaria of a nine-year-old boy in Rome at the turn of the twentieth century spurred several Italian physicians to carry out research aimed at solving the riddle of the cause of the disease. Ettore Marchiafava (1847–1935) and Augusto Celli (1857–1914), both from the Faculty of Medicine of Rome, believed that the pathogenic factor of malaria was a bacterium, *Bacillus malariae*. After careful studies, however, they agreed with Laveran and recognized the protozoan nature of the malarial parasite. They proposed the name *Plasmodium* for the protozoan identified and named by Laveran. In fact, in 1885, *Oscillaria* had been assigned to another organism, a "blue-green alga" (currently a cyanobacterium). Marchiafava and Celli also identified two species of *Plasmodium*, *P. falciparum* and *P. vivax*. Camillo Golgi (1843–1926), Professor of General Pathology at the University of Pavia, in northern Italy, was well-known worldwide for his research in the physiology of the nervous system (he was awarded the 1906 Nobel Prize in Physiology or Medicine, together with Santiago Ramón yCajal). Nevertheless, he was also deeply involved in malaria research. Indeed, residents of the agricultural areas along the Po River had a high risk of the disease, especially where there were extensive rice fields, such as the countryside around Pavia. Golgi made a notable contribution to malariology by relating the clinical sign of the fever episode with the schizogonic phase of the plasmodium, and by showing that the so-called tertian and quartan intermittent fevers are due to the presence in the blood of two different *Plasmodium* species (*P. malariae* and *P. vivax*), sometimes present together. Last, but certainly not least, was Giovanni Battista Grassi (1854–1925), the other main character in this story. Grassi (Fig. 1) graduated in Medicine at Pavia and, felt driven to become a zoologist because of his frequent contact with nature during his childhood and adolescence and because the University of Pavia was, at the time, the "sun of Italian biology", as Grassi himself used to say. After graduation, he worked at the Naples Zoological Station, founded by Anton Dohrn (1840–1909), and at the Messina oceanographic station of Nicolaus Kleinenberg (1842–1897). His training was completed at the University of Heidelberg under the guidance of two great scientists: Carl Gegenbaur (1826–1903), who reorganized comparative anatomy in Darwinian terms, and Otto Bütschli (1848–1920), one of the greatest experts on protozoans. At a very young age, Grassi became Professor of Zoology at Catania and was already famous for two extensive monographs, one on the Chaetognatha and the other on the vertebral column of fishes.

In 1888, Grassi began to study malaria in birds at Catania, in collaboration with the medical clinician Riccardo Feletti. In 1890, they published a monograph in which they described the malarial cycle in different species of birds, including owl, pigeon, and sparrow. It is here that we recognize Grassi's zoological approach to the problem, the constant physiognomy of Battista Grassi's method of investigation: different species of birds, belonging to different orders (Strigiformes, Columbiformes, and Passeriformes), are parasitized by different species of protozoans: only *Halteridium* in pigeons, but also *Proteosoma praecox* (*Haemoameba*) in the sparrow.

All this took place five years before Ross turned his attention also to the study of malaria in birds in India. Grassi was appointed Professor of Comparative Anatomy at La Sapienza, in Rome, in 1895. By then, his international reputation was established, not so much for his work on malaria in birds, but for his epochal contributions to entomology and biological oceanography concerning the life cycle of the common eel. In 1896, The Royal Society of London awarded him the prestigious Darwin medal, which, as specified by the zoologist Edwin Ray Lankester (1847–1929) in a letter announcing the award and now preserved in the Grassi Archive at La Sapienza, rewarded "those naturalists who are still in active work and especially doing work which has an important and direct role bearing on Mr. Darwin's own investigations & theory". Before Grassi, the prize had been awarded to Alfred Russell Wallace (1823–1913), Joseph Dalton Hooker (1817–1911), and Thomas Henry Huxley (1825–1895). In the opinion of English zoologists, therefore, the value of Battista Grassi as a naturalist was equal to that of the three greatest and most faithful friends of Charles Darwin.

In Rome, Grassi met the group of Roman malariologists, who convinced him of the validity of the transmission of the plasmodium via a hematophagous insect, a hypothesis he had until then considered doubtful. The problem was to identify the insect responsible for transmission. Grassi approached his research with the tools of the zoologist, namely knowledge of the systematics of the group and of the geographical distribution of the species. On the basis of the epidemiology of malaria and the distribution of mosquitoes present in the malarial zones, he focused his investigations on three species suspected of malarial transmission, *Anopheles claviger* (synonym *A. maculipennis*) and two *Culex* species (but not including the common *C. pipiens*), and he communicated this result to the Lincei Academy on September 19, 1898. On November 6, 1898, Grassi announced to the Lincei Academy that, with two colleagues, Drs. Bignami and Bastianelli, he had infected a volunteer by exposing him to the bite of these three mosquito species. Suspicion of the two *Culex* species faded immediately and the mosquitoes were acquitted of the crime of being vectors of the infection. On November 28, 1898, a formal note was sent to the academy and read in the academic session of December 4, 1898, where it was announced that a healthy man in a non-malarial zone had contracted tertian malaria after being bitten by an experimentally infected *Anopheles claviger*. The experimental phase ended on December 22 with a communication to the Lincei Academy that described the entire developmental cycle of the plasmodium in the body of *Anopheles claviger* and stated that it corresponded to what Ross had described for *Proteosoma* in *Culex pipiens* in the malarial cycle of birds. The experiment had been conducted with exceptional rigor; the *Anopheles* mosquitoes were raised in the laboratory beginning from the larval stages and starved until they had bitten a patient who had semilunar bodies in his blood—the only stage that could have developed into gametophytes and thus triggered the gonochoric cycle in the body of the mosquito. A healthy person was then exposed to the bite of these mosquitoes in a place protected from the introduction of other mosquito species.

The first issue of the Annales de l'Institut Pasteur of 1899 contained an article dated "Calcutta, 31 December 1898" by Major Ronald Ross and entitled "Du rôle des moustiques dans le paludisme". The insect responsible for transmission of the disease was indicated as "moustique d'une nouvelle espèce", just as in the note of 1897 it was indicated as a "grey" or "dappled winged" mosquito, which were absolutely invalid names for the Linnaean nomenclature. Ross was not a zoologist, however, and he completely lacked the tools of zoological systematics. At the margin of Ross's article in the issue of the Annales Pasteur, Grassi made many handwritten notes, including: "*non dice che fosse Anopheles*" (he doesn't say that it was *Anopheles*). On June 4, 1900, the Memoirs of the Royal Lincei Academy published an article by Grassi entitled "Studies on Malaria by a Zoologist". It consisted of 200 large-format pages that summarized the research Grassi had carried out from 1896 to 1899. In the article, Grassi defined himself as a zoologist.

In the crucial years of Roman research on the transmission of malaria (between 1897 and 1898), the English physician Edmonston Charles visited Grassi's laboratory in Via de Pretis and those of the other malariologists at the Santo Spirito Hospital. The Italian scientists, flattered by the interest of an English colleague in their studies, greeted him without suspicion of his motives. He then reported to Ross the information obtained. When the polemic about the priority of the discovery of the insect responsible for malarial transmission emerged, Ross felt obliged to make public the letters received from Charles. Charles wrote to Ross: "[…] I called on Dr. Manson before leaving London to get the latest news of what progress you had made in your work, in order to let the Italians know. They have been working in various directions this summer, but up till this week without being able to show any definite results. […] According to Grassi it would seem there are some fifty varieties of mosquitoes in Italy. Only six, however, seem to frequent these selected malarial positions. Besides the mosquitoes, the larvae were also brought up, and allowed to develop in Rome." The Roman malariologists eventually became mistrustful of the Englishman and did not tell him the truth about their advancements. In fact, by November 4, 1898, the suspects had already been limited to just three species and the acquittal of the two *Culex* species had already been decided. In another letter, dated November 19th, Charles wrote to Ross: "I went to his [Grassi] laboratory to try and get him to give you a few specimens of the different kinds of mosquito. I did this under the impression that he had completed his investigations. He told me, however, they were far from complete, and did not give me the specimens." Charles' initial impression was right.

By then, Grassi, Bignami, and Bastianelli had actually identified the malarial vector, and on December 4th, only two weeks after Charles' visit, they published their finding in the Reports of the Lincei Academy. Charles' letter then continues with an interesting sentence: "He [Grassi] spoke in the highest terms of praise of your work; he has your first report [the note of December 18, 1897 in the *British Medical Journal*], and told me to write to try and get your future reports at an early date for him."

On that date, therefore, the relationship between the two scientists was one of mutual respect, confirmed in a letter dated "Calcutta 5 February 1899" that Ross sent to Charles: "My dear Dr. Charles, very many thanks for your last letter with the translation of Grassi, Bignami, and Bastiamelli's note. This is good indeed. Pray give them my congratulations. [...] I thought that the grey mosquito is *Culex pipiens*, but was not quite sure. Of course there is a whole family of allied grey mosquitoes." This friendly and collaborative climate continued in the spring–summer of 1900, when Patrick Manson organized a crucial experiment to be conducted in an Italian malarial zone. A small building, in the style of English hunting huts, was designed and built in England and then assembled in Italy, in the Castelfusano pinewood on the hunting estate of the kings of Italy, which was near Ostia, not far from Rome. Two intrepid doctors, the Italian Luigi Sambon and the Englishman G.C. Low, both of the London School of Tropical Medicine, one of the most prestigious centers for the study of tropical diseases at the time, dwelled in this "mosquito proof" hut over the period of the summer–autumn fevers, which was also the period of maximum reproductive activity of the mosquitoes. The doctors, and their servants, remained free of malarial infection after a stay of three months. Bastianelli and Grassi followed the experiment, and the latter sent Manson a telegram dated September 13, 1900: "Assembled in British mosquito proof hut having versified (*sic*) [instead of "verified"] perfect health of experimenters among malaria stricken inhabitants. I greet Manson, who first formulated mosquito malaria theory."

At the end of 1900, Ross began a defamatory campaign against the three Italian researchers to claim priority of discovery of the mechanism of transmission of malaria, surely with the prospect of the Nobel prize in mind. He questioned the originality of Grassi's research, insisting that his indication that a "grey mosquito with dappled wings" was responsible for the transmission had guided Grassi in the identification of the vector. He even accused Grassi of fraud, on the basis of incorrect dating of the notes presented to the Lincei Academy, which instead had been precisely certified by the date of presentation in the Academy's public session. Grassi reacted harshly to these accusations, which he felt offended his honour as a scientist.

The dispute between Grassi and Ross about the priority of discovery has been interpreted as motivated by personal ambition, national pride, the desire for academic pre-eminence, and similar psychological and sociological explanations that have very little to do with science. In this regard, in 1998, Bynum wrote that the dispute "[...] is one of the least attractive episodes in the whole history of malariology," and this may be partly true; both scientists had strong personalities, and it was not easy to find a point of agreement. It might have been objected that, all in all, to have attributed a Linnaean name to the insect was a marginal part of the problem, but if this might have been justified in the nineteenth century it was no longer so at the threshold of the twentieth century. This is the unanimous judgment of science historians, who now attribute equal merit to both scientists.

The true meaning of the dispute, though, reflects the different ways of tackling problems in biological research, in this case, parasitological cycles. Grassi's approach was the one characteristic of zoological research: systematic, comparative, experimental. In contrast, Ross's approach was empirical and intuitive. Medicine in the 1800s, but also afterwards, up until recent times, was an empirical science. Not so zoology, which the Darwinian revolution approached with positivistic concreteness. For a post-Darwinian zoologist, the question of species was the focal point of the process: an animal remained undefined until it was placed in a context, no longer merely a classificatory context, but also an evolutionary one. For Grassi, nomenclatural meticulousness was almost an obsession. The second tool that guaranteed Grassi's success was the comparative approach, which he thought was the main route to arrive at the answer of the malaria question. In his research, he excluded species that could not be malarial vectors and unequivocally identified *Anopheles claviger* (syn. *A. maculipennis*) as the only vector of malaria in Italy. For Grassi, parasitology was a zoological science; it was the application of Darwinism to pathology.

Grassi also had his dogma: "There is no malaria without *Anopheles*". The opposite, however, was not always the case: he noticed that there were areas where *Anopheles* was abundant but malaria was absent. An initial hypothesis, expressed in the second edition of his article, was to relate this phenomenon to the thermophily of the *Anopheles* mosquito. In areas with cold nights, the mosquito did not fly and bite humans, but stayed in the warmer stalls to bite livestock.

After World War I, malaria flared up with renewed vigour. Mortality from malaria had rapidly decreased in Italy starting from 1898, after discovery of the vector and institution of zooprophylactic activity, as well as the free distribution of quinine. As a result, in 1915, the number of deaths had decreased from 600 per million inhabitants to less than 50; however, it then increased to 320 per million in 1919 (Fig. 3). Resuming his research, Grassi turned again to the problem of "anophelism without malaria". He identified three localities with a typical malarial environment, all infested by *Anopheles maculipennis* but not affected by malaria. In 1921, he demonstrated that "there is certainly a biological race of *Anopheles* mosquitoes that does not bite man." A year after Grassi's death in 1925, one of his pupils showed, on the basis of these observations, that there are six species of *Anopheles* in the *maculipennis* complex, which are indistinguishable except for their egg morphology. Of these six "new" species, *Anopheles labranchiae* and *A. sacarovi,* present in highly malarial zones, mainly bite humans, while the typical form of *A. maculipennis* only bites animals. Another species, *A. messae*, mostly bites animals, but sometimes also humans. These findings demonstrated the importance of precise systematic identification of the vector in antimalarial prophylaxis; rough identifications like "grey mosquito" or "dappled winged mosquito" could not suffice. It is interesting that complexes of *Anopheles gambiae* and *A. arabensis*, the two species most responsible for malarial transmission in the world, also show anthropophily and zoophily, characterizing different populations of the two species.

Malaria was eradicated in Italy through systematic control of the insect vector. This was conducted on two fronts: reclamation of marshy environments, and direct biological and chemical control of the mosquito. Although the former involved hydraulic engineering projects, the latter mainly required the work of zoologists, direct and indirect pupils of Grassi and Bignami who continued their work.

Much of the above source of information is based on Capanna (2012), Spielman & D'Antonio (2001) and Harrison (1978).

Malaria still claims many victims throughout the world—far too many in countries, like Italy at the end of the 1800s, that cannot proceed on the road to development until the disease has disappeared. Is the Italian model applicable to such countries? The problem is certainly much more complex; the vastness of the areas affected, sometimes whole continents or subcontinents, and the ecological and socio-economic situations hinder antimalarial campaigns. New strategies of vaccination or of genetic engineering aimed at creating transgenic *Anopheles* mosquitoes are being attempted. Therefore, there arises once again the need for genetic characterization of *Anopheles* populations—a task requiring that zoologists and geneticists work together.

• • •

14

Was Dr Ross's 'Dapple-Winged Mosquito' an *'Anopheles stephensi'*: A Fine and Isolated Verisimilitude – *A Taxonomic Analysis of Penetralium of Mystery?*

"Imagination is
the highest form of Research."
– (Albert Einstein)

On the advice of Dr Patrick Manson, who had earlier proved role of mosquito (*Culex pipiens*) in the transmission of worms of *Filariasis* [*Filaria (Wuchereria) bancrofti*] while working in China, in 1880, Dr Ronald Ross converged his attention from cockroaches, fishes etc. etc. to the mosquitoes. He, however, did not know even the elementary knowledge of mosquitoes which had already appeared as early as in the classics like *Systema Naturae* (Limmaeus, 1758) and *Reise nach Palgstina* (Gadebusch 1762 *in* Knight, 1972). Many other zoologists had also catalogued mosquitoes during the 18th century such as for example De Geer, Fabricius, Forskål, Fourcroy, Muller, Olivier, Pallas, Poiret, Rossi, and Schrank (*cf.* Knight, 1972). Many workers from the early 19th century were further added to this list. Although a few mosquito descriptions were published earlier, the official beginning of mosquito systematics is January 1, 1758, the date established by the International Code of Zoological Nomenclature as marking the beginning of the consistent general application of binominal nomenclature in Zoology. The *Anopheles* (?) species Ross had spotted in Sigur Ghat was not taxonomically described yet; it had to wait until five-six more years before the same was brought on record Liston in 1901 as *Anopheles stephensi* from Ellichpur near Berars in Deccan of India (14.1). Obviously, therefore Ross had but picked up the name *An. stephensi*, at a much later date, to associate it with his Sirgur Ghat's 'Dappled Winged Mosquito', most arbitrarily though.

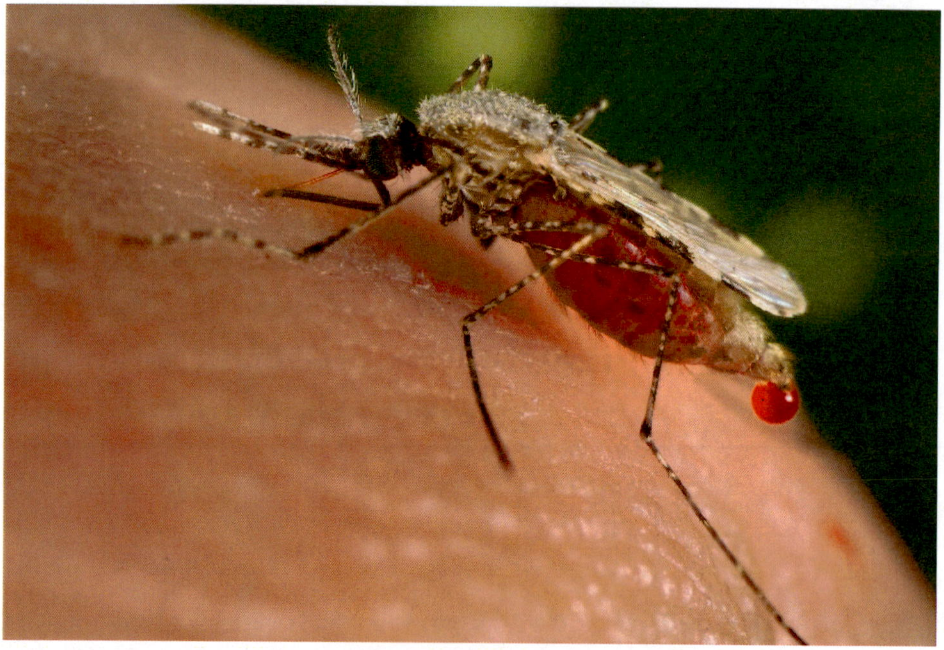

Fig. 14.1. *Anopheles stephensi* Liston, 1901 (*Source:* Wikipedia)

Mosquitoes are slender, long-legged insects that are easily recognized by their long proboscis and the presence of scales on most parts of the body. Larvae are distinguished from other aquatic insects by the absence of legs, the presence of a distinct head bearing mouth brushes and antennae, a bulbous thorax that is wider than the head and abdomen, posterior anal papillae and either a pair of respiratory openings (subfamily Anophelinae) or an elongate siphon (subfamily Culicinae) borne near the end of the abdomen. Mosquitoes are usually, and most reliably, identified as mature (fourth-instar) larvae and adults; sometimes pupae may be an additional help in identification. Males are especially needed to distinguish many species because the females of various generic-level taxa are remarkably similar in habitus (Fig. 14.2). Whereas the taxonomy of numerous apparently closely related species is confounded by overt similarity, the limits of many supraspecific groups are clouded by morphological diversity. Some genera include diverse elements of indefinite affinities that will inevitably be recognized as separate monophyletic lines once they are thoroughly studied (Harbach & Kitching, 1998; Reinert et al., 2006).

The immature stages of mosquitoes occupy a spectrum of aquatic environments. They occur primarily in temporary or permanent bodies of ground water, but a large number of species occupy leaf axils, tree-holes, rock-holes, crab-holes, bamboo internodes, bromeliads and aroids, fruit shells and husks, fallen leaves and spathes, flower bracts, snail shells and pitcher plants. Some utilize artificial containers as well as the normal ground-water habitats.

The majority of larvae feed on suspended particulate matter and microorganisms that they extract from the water with filamentous mouth brushes. Other species are obligatory or facultative predators that capture and feed largely on the immature stages of other mosquitoes by means of modified mouth-brushes or grasping mandibles or maxillae. Some larvae resort to scavenging or cannibalism when food is scarce. The larvae of most mosquitoes obtain oxygen from the atmosphere by coming to the water surface. All species of Mansonia and Coquillettidia and some species of Mimomyia obtain oxygen from the air vessels of aquatic plants, which they pierce with a specialized siphon. Aedeomyia species apparently use their enlarged antennae for respiration. Some species have greatly enlarged anal papillae that are well supplied with tracheae, and these species seldom come to the surface and probably obtain dissolved oxygen from the water. In view of the ecological diversity of mosquitoes and the broad range of aquatic habitats they occupy, they can be used as environmental indicators on a variety of scales. As an example, the immature stages of numerous mosquito species inhabit a variety of tropical plants (phytotelmata), which makes them good indicators of the health of forests Since mosquitoes are delicate insects, they are always found where the air is relatively cool and the humidity is high. Many species live within a few meters of the ground whereas many sylvan species occur primarily in forest canopy. Vertical distribution is largely dependent on feeding preferences. All males and the females of many species feed exclusively on plant liquids, including nectar, honeydew, fruit juices and exudates. Females of numerous species feed on the blood of living animals, but some that are normally hematophagous may produce eggs without a blood meal. Warm-blooded vertebrates are a common source of blood for most species, but many species also attack cold-blooded animals such as snakes, turtles, toads, frogs and other insects, including nymphal cicadas, lepidopterous larvae and mantids. The time of flight and feeding activity is usually quite specific for most species. Some species are active at night (nocturnal) or twilight (crepuscular) whereas others are active during the daylight hours (diurnal). Despite our current knowledge of mosquito biology, practically nothing is known about the specific bionomics of most species.

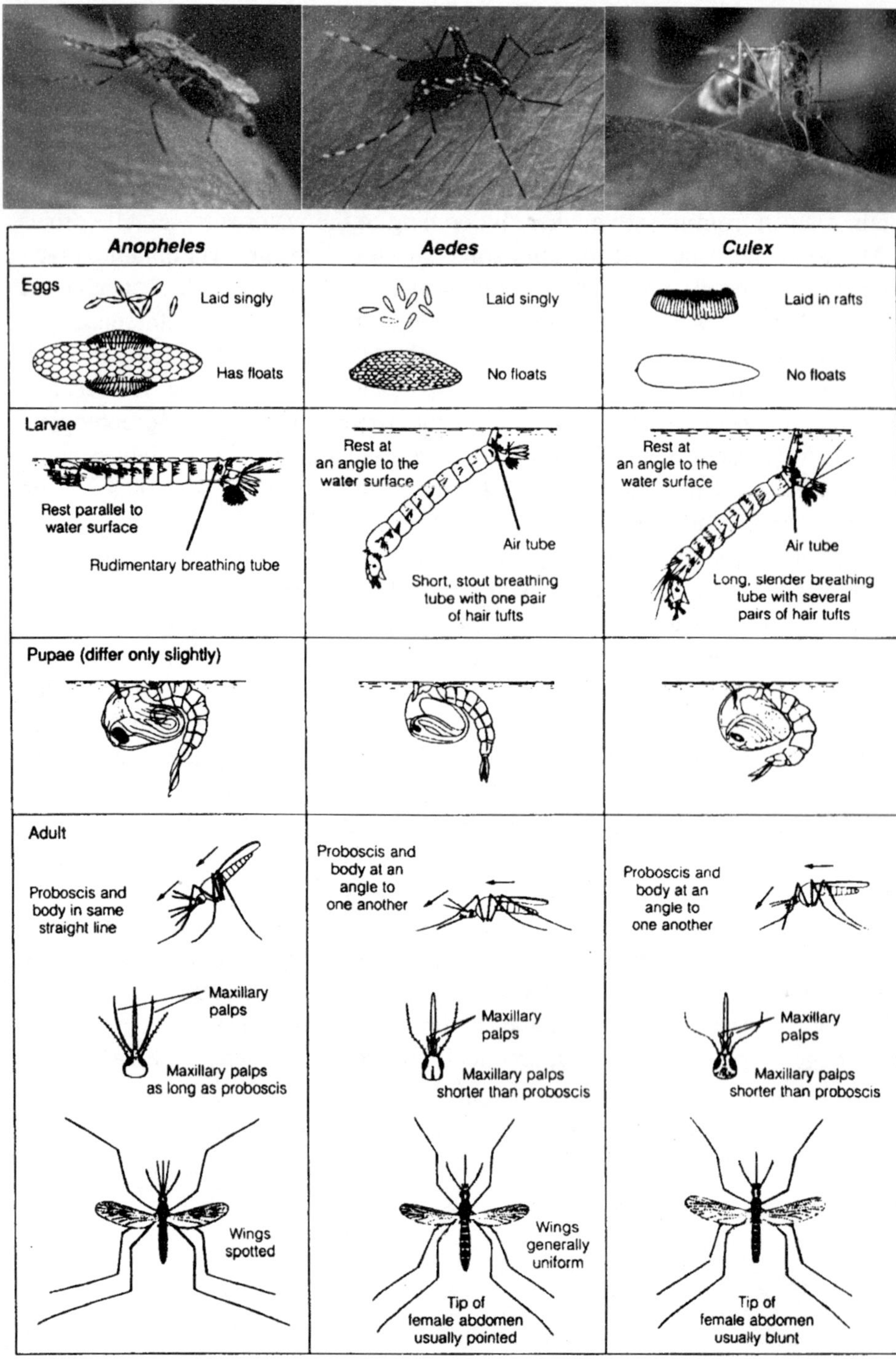

Fig. 14.2. Some of the main characteristics for differentiating *Anopheles*, *Aedes* and *Culex* mosquitoes.

Many species of mosquitoes are pests or vectors of pathogens that cause disease in humans and domesticated animals. The pathogens transmitted by mosquitoes include viruses (arboviruses), filarial worms (helminths) and protozoa. Fewer than 150 species, largely confined to genera Anopheles, Aedes (traditional broad sense) and Culex, are the indirect cause of more morbidity and mortality among humans than any other group of organisms. Despite their medical importance and long history of study, the taxonomy of mosquitoes is far from complete and the existing system of classification is not entirely natural.

Anopheles mosquitoes, in general, are strikingly different from all the Culicinae mosquitoes in having clearly discernible white-spotted wings and banded maxillary palps, besides of course many other distinctive features, especially on genitalia (Fig. 14.3a,b).

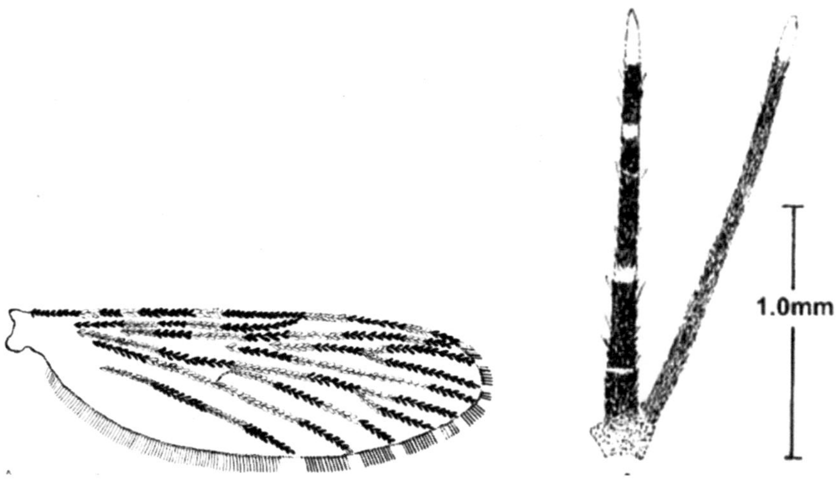

(a) **(b)**

Fig. 14.3a,b: An *Anopheles* mosquito's (a) wing venation with white markings, (b) white-banded palpi.

For a layman, among all the Indian anophelines, *An, stephensi* is rather more characteristically distinguishable from the rest of species by a combination of the following characters: Palpi apical segment all pale, forming, with apex of next segment, a broad pale apical band; a broad apical and basal pale band at 3-4, the dark intervening area narrower than either band; a narrow pale band at 2-3 and some spots formed by patches of white or pale scales on dorsum of segment 3; Dorsum of thorax with a conspicuous mat of white scales, and legs with femur, tibia and tarsi characteristically white-spotted (Fig. 14.4).

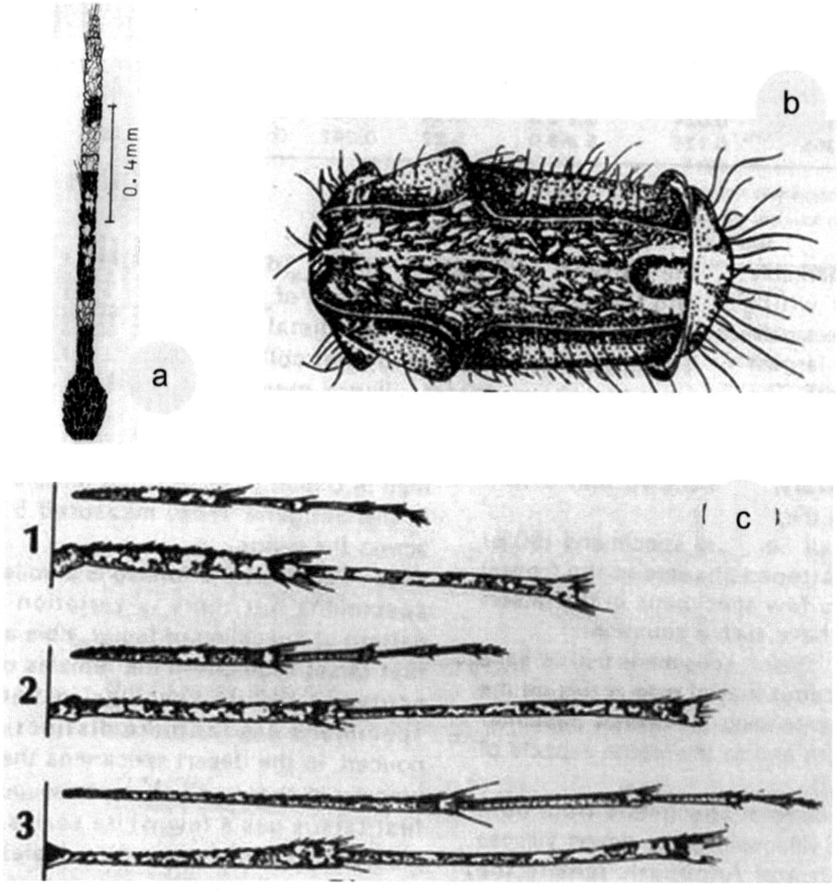

Fig. 14.4. *Anopheles stephensi*: (a) palpi, (b) Thorax, and (c) Legs. (*Source:* Tyagi *et al.*, 1991.

More specifically, especially with reference to a humongous issue related to the identification of the species spotted by Ross first in Sirgur Ghat and subsequently, and also for the last time, before his pathbreaking Medical discovery in August, 1897, the mosquito, *An. stephensi*, is described (Christophers, 1933), as follows:

ADULT ♀.-Size medium (length of wing 2-5-4.0 mm.).

Head : scales of usual type, with a well-marked pale vertical area; vertical chaetae white, with a cluster of about nine flattened chaetae forming well-marked frontal tuft; ocular scales anteriorly forming about a single line. Antennae with small pale scales on torus; white scales on first 2-4 flagellar segments. Palpi moderately thick, cylindrical, smooth except towards base; apical segment half-length or more of preceding segment (index 0.58); apical segment all pale, forming, with apex of next segment, a broad pale apical band; a broad apical and basal pale band at 3-4, the dark intervening area narrower than either band; a narrow pale band at 2-3 and some spots formed by patches of white or pale scales on dorsum of segment 3, these, though sometimes absent, being very characteristic of the species.

Pharynx: filament of cones rather broad, flat, short, with lateral spicules and fimbriated ends ; crest broad, with double row of spines, bifid in posterior view ; rods with swollen base, tapering to rod-like terminal portion, simple.

Thorax usually with a few pale scales on *apn*. Mesonotum brown, fossae and lateral areas a little darker than median area: median area thickly covered with pale narrow scales (str. about 5-6), forming lateral tufts of white scales, with dark scales beneath on promontory; fossae with scattered scales and a heavy line of scales on lateral area in front of wing-root, some scales with 8-9 str. Pleurae without obvious scales; spir. hairs 2-3; upper mesep. 5-6.

Wing as in Fig. ..., usually with prehumeral dark acc. costal spot undivided with white scales anteriorly; an interruption present or not at inner end of inner costal spot; marking on vein 1 in middle costal spot very variable or with a small dark spot also at inner end, or both small dark spots absent; dark spots on stem 2, 2.1, base 3, and some others very variable in extent.

Legs with front femora only very slightly swollen in basal portion; femora and tibia? speckled as shown in figure, hind femora and tibiae usually pale on inner aspect; first tarsal segment with some spots; others dark, with apical and basal bands on joints l-2 and 2-3 on fore legs, narrower apical banding on mid- and hind legs, the mid-tarsus often mainly dark.

Abdomen with narrow scales on tergites II-VIII, increasing in extent towards posterior segments, which are rather heavily covered with narrow oblanceolate scales mixed with pale hairs, VIII with dark scales at lateral posterior angles; some dark scales apically in mid-line on venter of VII; cerci with dark scales.

ADULT ♂. -In general as in ♀.

Antennae: without scales on first flagellar segment. Palpi ornamented; marginal hairs forming about a single row above and a double row ventrally. Coxites with numerous scales.

Hypopygium: harpago with apical hair somewhat longer than club; a hair about half its length or less between this and club, and from l-3 longish hairs internal to apical hair, with numerous rather large smaller hairs about apes. Phallosome between half and 4 coxite, with 5-6 leaflets on each side;

practically all the leaflets broad, blade-like, with marked serrations along straighter edge; the largest about half length of phallosome.

PUPA

Paddle: spines on posterior half or so of external border fine, pointed, giving place to hairs on the posterior border, which do not extend beyond the paddle-hair ; paddle hair long, hooked.

Spine: (VIII) rather long, more than half length segment; (V-VII) about half length of segment; (IV) about half length of preceding spine, usually blunt (sometimes pointed and rather longer); (III) small, about i length preceding; (I) minute, unchitinized.

Hair B: (V-VII) about 2/3rd length of segment; about 3 br., leash-like.

Hair C: (V-VII) simple, longer than segment; (IV) 2-3 br., leash-like. Hair 4 on segs. III-VI rather long, approaching half segment, simple.

LARVA.

Clypeal hairs: simple, *ic* sometimes very finely frayed ; *oc* about 2/3rd length of inner; *pc* about the same length as *oc*. Antenna with hair arising about 1/3rd length of antenna from base; terminal hair split about middle into 2-4 br. *Maxillary palp* with the cone slightly bifid at extreme tip. *Mentum* with four teeth on either side of median tooth, the first of these very short and rounded, the first three more or less equidistant.

Shoulder hairs: inner and middle hairs arising from separate basal tubercles, variable in amount of development, but tubercle of middle hair always better developed than that of inner; both hairs feathered, inner somewhat shorter than middle. *Metathoracic hair no. 1* like ordinary hair, usually with three branches (sometimes 4-5). Hair no. 5 on prothorax long and feathered, its distal end drawn out and not having truncated appearance. *Pleural hairs* as given under the group; *dp*l with 3-4 br. Tubercles moderately large, that on 1 and 3 with the projection sharp and spine-like, that on 2 blunt and bifid. **Palmate hairs** well developed on III-VII; hair no. 1 on I not developed as palmate hair, that on II poorly developed. Leaflets uniformly pigmented, broad; filament short, broad at base, about 3 length of blade; may be longer in specimens from northwestern areas. *Lateral hairs* on IV-VI split some distance from base into 3 br., sometimes 4 or 5 on IV and 4 on VI ; on VII very short, with 3-5 br. *Tergal plates* rather small. *spc* poorly developed ; *mps* very broad anteriorly and touching chitinization. Pecten with 3-5 long and 8-11 short processes, all serrated on basal half. *ps* with 5-6 br. Tail-hooks poorly developed.

Finally, Christophers conclusively added about the identification as: ***There is no species with which this can well be confused; it is the only Indian species with broad thoracic scaling and a double broad pale band on the palpi in which the hind legs are not conspicuously marked with white.***

Ross knew nothing of the above.

The second time Ross had viewed the 'Dappled Winged" mosquito, about a dozen, he had found some difference in size and felt that the Secunderabad specimens were slightly bigger in size if compared with the Sirgur one. He was careful to keep the dissected specimens (plesiotype) for taxonomic confirmation, which however was hardly carried out in writing! He wrote to Manson on 12th May, 1897, about his first catch in Sigur Ghat on the 7th May, 1897, the 'C' type, with characteristic 'dapple wings', and described its features as follows:

> "... caught a mosquito full of blood (in my Kalhutti) rest house. This mosquito was small and brown, with wings striped brown and white."

Later on he corrected himself and replaced 'stripes' with 'spots' on the wing.

Around 1898, largely due to publication of research findings by the Italians, Ross better thought to assign a name to his 'C-type' mosquito which he called likely to be *Anopheles stephensi.* In his *Memoirs,* Ross (1923) carefully redefined his mosquito in the following way:

> "... which bite at night, have spotted wings, and sit on a wall with tail sloped away from it. Their eggs are boat-shaped , laid singly, and often to lie together touching each other end to end so as to form equilateral triangles. Their larvae have no breathing tube and float flat on the surface of the water like sticks"

If one carefully goes through the entire account of events surrounding the discovery of 'dappled winged mosquito' by Ross, along with his various descriptions of the species, the conclusion can be safely drawn that Ross was never sure about the species being *An. stephensi.* Scientifically, too, whatever Ross might have inferred from his observations about the species, the mosquito in question did not at all fit into the definition of the species, *An. stephensi*. And, if all the potentially commensurate reasoning is set aside, it would be questioned whether the 'dappled winged mosquito' could be actually the race or subspecies, *mysorensis,* which really is quite common in certain parts of the rural environment of southern India.

• • •

15

Kishori Mohan Bandhopadhyay: *The Unsung Hero behind Ronald Ross's World Famous Discovery*

"No man will make a great leader who wants to do it all himself or get all the credit for doing it."

– (Andrew Carnegie)

Britishers, during the colonial period, started the 'orderly' system in service, especially the armed forces. An orderly is an attendant officially deputed to work either at the house of an officer; the higher rank officer were entitled more orderlies and lower rank officers the minimum number. Dr Ross, too, was allotted one or two such attendants, to work either as '*Mali*' (Gardener) or as the one to set in order his daily uniform, from shoes to belt to cap, for the office. If the officer concerned did not mess at the regiment he could also use one orderly as his '*Maharaj*' (Cook). In Office the British, and other Europeans, invariably had a battery of native assistants at different levels. In fact, the European officers and scholars were to a great extent dependent on the expert assistance of these Indian talents to achieve success [e.g., Radhanath Sikdar (1813–1870), the great Indian Mathematician employed as 'Computer' first in the Survey of India and later as the 'Chief Computer' as well as Superintendent for Meteorology Department who discovered the exact height of the Mt. Everest as 29,000 feet, but his boss Colonel Andrew Scott Waugh, Surveyor General of India treacherously increased the height by arbitrary two feet, making it 29002 ft., and the formally announced it in March 1856, thus taking all the credit and hogging all the limelight in the heat of tat grand discovery!]. But they seldom gave the natives any credit due to them for their efforts. The story of Kishori Mohan Bandhopadhyay is an eye opener (Fig. 15.1). Dr Ross, honoured with the Nobel Prize for the discovery of the malaria parasite in mosquito completely overshadowed him without any commendable mention anywhere (there is, but a passing reference for a capable laboratory assistant of the caliber of Mohan Kishori Bandhopadhyay in his '*Memoirs*'), either in his scientific papers, daily diary, correspondence with Dr Manson, or books. This is ridiculous, bizarre and unacceptable since he had himself not been able to tolerate Dr Grassi's (very minor, if at all necessary) missing on giving him the due credit in his paper presented to *Lancei* in 1898. He then waged a war of words, defamation and threat to drag Dr Grassi in the international tribunal.

When Dr Ross won the Nobel Prize many eyebrows were raised, but the most striking was from India – the country which provided him all the opportunities to accomplish his epochal discovery and win the highest decoration in science. The maximum damage was caused when he read out his Nobel Prize Lecture before the Swedish Committee without even mentioning the name of the main anchor of his new finding on which was based his entire glory. In order to express their consternation Biesel & Boete (2013) commented, "Even though Ross was the sole recipient of the Nobel Prize he was by far not the only one who hypothesized that mosquitoes transmit malaria. Amitava Ghosh (2001) has mentioned repeatedly in his award winning novel "*The Calcutta Chromosome: A Novel of Fevers, Delirium & Discovery*", based on the life and works of Dr Ronald Ross, as to how he had exploited native workers in his quest to find the cause of malaria without giving them due honour.

The importance of his Indian research assistant Kishori Mohan Bandhopadhyay is acknowledged by many. The fact that he was not a co-recipient of the prize, but merely received a Gold Medal, is read by many as an effect of colonial relations." However the alternative recognition of Kishori Mohan was far from being commensurate. It is important here, therefore, that professional life of Kishori Mohan Bandhopadhyay is portrayed in the right context – to exhibit that Dr Ross's discovery was almost impossible without the dedicated contribution of his laboratory assistant.

Fig. 15.1. Kishori Mohan Bandhopadhyay (*Source:* Dutta, 2009)

Kishori Mohan Bandhopadhyay was born in Entally area of Kolkata at his maternal grand parents' home. On paternal side, the family moved from Bengal Jessore (now in Bangladesh) to Calcutta in quest of better vocation like many other Bengali families during colonial times. They settled in Panihati, 24 Pargana near Kolkata. His father Nanilal Bandhopadhyay was a teacher in Sanskrit and Persian literature and languages.

His family was highly educated and two elder brothers, Lalmohan and Haridas, were teachers in Bengali and Sanskrit, respectively. Till date his descendants and other family members live there in a house called '*Nilambati*' in Panihati area in the Kolkata metropolis.

Kishori Mohan started his education at Panihati's vernacular school where his father and brothers were teachers. Then he was sent to his maternal grandfather's house in Calcutta where he completed entrance examination of University of Calcutta and graduated in science from the Presidency College. After completing graduation, he joined as a science teacher in an English medium school in pursuance of his family tradition.

In February, 1898, when Dr Ross advertised for enrolling an assistant he submitted his candidature to him. The same month an interview was held and Dr Ross was greatly impressed by his overall biological knowledge. Consequently, out of the many applicants Kishori Mohan was recruited to help in his malariological explorations. He joined as the Laboratory Assistant to Dr Ronald Ross at the Calcutta Presidency General Hospital (Fig. 15.2).

Fig. 15.2. Dr Ross's laboratory when Kishori Mohan permanently had a seat to work. (*Source:* Dutta, 2009).

Dr Ronald Ross was noted to be eccentric and egocentric, described as an "impulsive man". His professional life appeared to be in constant feud with his students, colleagues and fellow scientists. His personal vendetta with G.B. Gracie became a legendary tale of science. He was openly envious of his mentor Patrick Manson's affluence from private practices. This was largely due to his own ineptitude to compete with other physicians. He hardly had good ties with the administration of Liverpool School of Tropical Medicine, constantly whining about being underpaid. He resigned twice, and was eventually discharged without any pension. All these evidences indicate that it was very natural for Ross to deliberately drop out Kishori Mohan's contribution in his research work. During the colonial period it was difficult for Europeans to give equal weightage to an Indian.

When Ronald Ross received the Nobel Prize in 1902, to honour the contribution of Kishori Mohan, Upendranath Brahmachari, Acharya Jagadish Chandra Bose, Brajendra Nath Seal, Sivanath Sastri, Surendranath Banerjee, and Acharya Prafulla Chandra Ray requested Lord Curzon to give recognition to Bandhopadhyay. Lord Curzon took the initiative and Bandhopadhyay was given King Edward VII's Gold Medal in 1903 during the Delhi Durbar by the Duke of Connaught.

Fig. 15.3. Kishori Mohan being honoured by countrymen. (*Source:* Dutta, 2009).

On his return from Delhi after receiving King Edward VII's Gold Medal, Kishori Mohan Bandhopadhyay was felicitated at a function in the Senate Hall of Calcutta (now demolished) by Bengali scientists and doctors (Fig. 15.3). The function was presided over by Chittaranjan Das. The famous physician Upendranath Brahmachari delivered the main lecture eulogizing his work. Other speakers included Brajendra Nath Seal, Shivnath Sastri and Bidhan Chandra Roy, who later became the Chief Minister of West Bengal. Even though he was honoured by his own countrymen, Bandhopadhyay was disappointed when Ross published his memoirs, With a Full Account of the Great Malaria Problem and Its Solution in 1923 without mentioning his name. He eventually refused to meet Ross, who revisited the Presidency General Hospital in 1927. In a few places he had mentioned his assistants, but not by name. Later, Bandhopadhyay became a social worker. He started social campaigns in villages in Bengal for eradication of malaria. Kishori Mohan Bandhopadhyay started touring the villages in Bengal with his Magic Lantern and slides depicting parasites, mosquitoes, patients and steps for prevention of the disease. He organized slide shows in order to educate the village folk about malaria or Sheet Jwar, as they called it. The black-and-white slides were prepared by his photographer friend Lakshminarayan Gangopadhyay who resided in Uttarpara, Hooghly across the Ganges river, and who was the first roving photo-artist in India. The black- and-white plates were coloured by Lakshminarayan with transparent dyes so that the slides became understandable to the villagers. The images on glass were generally projected on a white cloth hung on the outside wall of a villager's house.

His experience based on visits to various villages in Bengal during the anti-malaria campaigns as well as in connection with his work relating to the freedom movement, convinced Kishori Mohan Bandhopadhyay that the poor and lower middle class people need credit from an alternative source to avoid the stranglehold of moneylenders. Whatever savings he had after his success post the Gold Medal, he pulled them together and sold off the gold ornaments of his wife to start Panihati Co-operative Bank with a few like-minded friends in 1927. The bank was opened on the same road on which he had his ancestral home 'Nilambati'; the road at that time was called Banerjee Para Road, now renamed as Ramchand Ghat Road. After having spent whatever he had for the co-operative bank, Kishori Mohan was almost penniless. His social work also suffered. He was, in those days, a one- man NGO. The bank was a great success in ameliorating the conditions of the poverty-stricken villagers in and around Panihati, and is still running. He joined politics and later campaigned for Bidhan Chandra Roy, who was his friend and well-wisher.

In 1929, Bandhopadhyay suffered from meningitis. Since no medical arrived timely, he succumbed to the infirmity and died in July 1929. The Panihati municipality has a street named Kishori Mohan Banerjee Road.

• • •

16

'The Nobel Prize' Factor: The Race for Supremacy and Priority in Malaria Research – *Friends Turning Foes*

"It is excellent
To have a giant's strength;
but it is tyrannous
To use it like a giant."

Never in the history was such a war fought or even imagined to be fought! It was like a world war, with a difference, though: the enemy was a tiny and feeble-bodied insect - the mosquito, still not quite known for its role in bringing deadly infections to man but had been always looked upon with draconian suspicion and great scientific interest alike. For this tiny, feeble-bodied mosquito was soon going to occupy the central position of all the world's strongest minds' attention and discourses unmatched in the annals of science of any other field. There have been many a great mosquito-related battle but none was fought like the battle of 'malaria-mosquito' interdependence, in the last quarter of nineteenth century! The tiny mosquito assumed such a high importance that the conflict engulfed more countries into its bitter folds than probably the real World War One nearly a quarter century later! But, even this 'great malaria- mosquito war' was fought in a military style. This was mankind's first world war against another species. But this, too, was an intra-specific war, between men of the subject, and with mosquito playing the pivotal role in the entire tumultuous saga of primacy and priority over the discovery of malaria-mosquito connection.

The 'Great Malaria Mosquito War' erupted with the first glimmer of proclamation on 20th August, 1897, by Dr. Ronald Ross, whom many claim as an unlikely, if not reluctant, scientist or malariologist, that mosquitoes transmit malaria. Although he could not scientifically recognize the species in the lack of any technical training in the subject of taxonomy and identification of different types of mosquitoes, he was able to distinguish among various different mosquito types on their colour patterns; 'grey' or 'bar-backed' mosquitoes (one of which was perhaps *Culex quinquefasciatus*, the common house mosquito; his 'A' type), 'brindled' mosquitoes (perhaps *Aedes aegypti*, the Yellow Fever mosquito; his 'B' type) and the big brown mosquito with striped (actually spotted) wings which he later called as the dapple-wing mosquito (which was in fact an anopheline mosquito belonging to the genus *Anopheles*, members of which are capable to transmit human malaria parasites and pass them to people; whether it was *An. stephensi* can only be conjectured). For this reason only, Dr. Grassi, considered amongst medical zoology fraternity a preeminent scientist with enormous experience in malaria/mosquito-related subjects, regarded Ross, who was completely unaware of mosquito taxonomy, as an amateur and an interloper. In the face of the background of the announcement of establishment of the prestigious Nobel Prize an year before in 1896, endowing in prize money for scientific discoveries the largest monetary payments ever, the first seeds of 'The great malaria mosquito war' or the war of giants were seemed to have been sown. This further mushroomed in a more vicious way in years to come, and for the rest of their lives, both Ross and Grassi, the former in particular, were horn-locked hurling insinuations of all types on each other with a view to secure priority or supremacy in several discoveries!

During the late 1870s and early 1880s, mosquito had begun to attract much attention of tropical disease researchers and Manson's assertion of mosquito's role in transmitting human lymphatic filariasis worked to further ignite subject-ebullience at a fevered pace all over the world. All the three serious diseases of the day, *i.e.*, malaria, filariasis and Yellow Fever, were being looked into with a microscopic precision. Still, the attraction which malaria commissioned was unparalleled in the history of vector-borne tropical diseases around the globe. Malaria always asserted some claim to be special. Widespread and debilitating, malaria - the most common of all tropical diseases - was of all time great interest, as scientists in France, Italy, Germany, United States, England and India vied in their independent and isolated pursuit with a rather nationalist fervor for research excellence and discovery claims. The struggle was no less intense than the Europeans' incessant voyages in 14th-15th centuries to discover the New World or 'the 'golden bird' that India had once been, and it foreshadowed future scientific contests over the cholera, polio or smallpox vaccine and the atom (Spielman & D'Antonio, 2001). Most of the names in circulation were those of much acclaimed and seasoned scientists; the Italian genius Professor Giovanni Battista Grassi, commanding a team of as dedicated and qualified researchers as Dr Amico Bignami, Dr Marchiafava and Dr Camillo Golgi, was already a hugely respected scientist with a unique knowledge of both medicine and Zoology, and a background of several years of mosquito-malaria research; Dr Patrick Manson, a true genius and creative mind always perceiving connections and resemblances between disparate phenomena through his sound and logical imaginations, was an undisputable authority on tropical diseases especially after his discovery in China about the role of mosquito in lymphatic filariasis transmission (*Filaria=Wuchereria bancrofti*); Dr Charles Luis Alphonse Laveran, though unimportantly dispatched to Algeria, in 1878, on an official assignment to help save the lives of French troops from some kind of quaint infection then unknown to the mankind. Providentially pitched into investigating malaria, however he discovered the first human malaria parasite, the crescent stage or the sexual bodies, belonging to *Plasmodium falciparum,* and also other asexual stages belonging to *P. malariae.* Unfortunately, he published results of his pathbreaking investigation in a rather non-descript French journal. In contrast to all these and many others who worked on malaria and mosquitoes, Dr Ronald Ross, though egoistical and ambitious, like Laveran, and often regarded as an unlikely, if not reluctant, scientist, on one hand, and an ever invigorated, inexhaustible, self-taught and obsessed malariologist, on the other, was in fact a genius *par excellence* who was ultimately credited for unraveling the long time baffling secrets of malaria parasite's journey outside human body in a mosquito!

All these scientists regardless of their academic and professional background had one thing in common; they were endowed with the seemingly limitless energy, patience and optimism that are indispensable traits of discoverers. ln this different kind of war game the enemy was unknown and the actors played their part of roles in all directions with a phlebotomist's precision with utmost caution, hoping to strike the bull's eye first amongst all. While Manson's pursuit of the filarial worm between late 1870s and early 1880s cleared all doubts and dark points pertaining to mosquito's role in effecting transmission of the filarial worm and ignited equally subtle minds around the globe, scientists particularly in Europe and North America competed with nostalgic resolution and pace to win the ultimate race to scientific discovery, which was certainly like hitting the jackpot.

All kinds of means were adopted to press in the claim of primacy; advance reports of 'successful' experimentations were regularly advertised in the premier dailies, scientists indulged in verbal and written duels to forward first their 'original' claim to secure credit, some scientists, if not all, lobbied not only in the world of science but in other public or social realms as well to muster in support to win the all-important victorious vote in the event of a contest for priority over discovery or when it mattered most, and, worst of all, a few scientists clandestinely spied over others' progress in work! Skepticism and doubt loomed large even though scientists, in order to understand the enigmatic matters of malaria a bit more better, logically involved in exchanging written, published or unpublished records, including microscopic slides. Patriotic appeals for increased funding in research became intense and nationalistic sense of competition took the uppermost tone in expressing solidarity to the home economy and health of the countrymen. Notwithstanding all this precarious environment, the scientific fraternity, comradeship and cooperation amongst scientists of various different countries too continued, though not always unscathed, as illustrated through the scores of correspondence made by Manson, Grassi and Ross in the midst of the year of discovery, *i.e.*, 1897.

Destined, as if it were, to effervesce an already ebullient quest for malaria mystery through mosquito, the announcement of setting up of the Nobel Prize in Medicine and Physiology (as a part of a series of world prizes instituted in 1896 following the death will of the dynamite king Alfred Nobel, a Swede who had a great benevolent heart and appreciation for scientists' remarkable contribution in making a better world to live in), with the largest monetary payments ever for scientific discoveries, apparently ignited every mind around genuinely engaged in deciphering malaria-enigma and to proclaim first to reach the ever awaited conclusion! The contest eventually narrowed down to two scientists of contrasting characters, Grassi and Ross. Professor Grassi was a zoologist of international repute and he had been seriously toiling to prove a malaria-mosquito connection for several years. In contrast, Dr. Ross, in the words of Grassi, was an interloper and an amateur who could not distinguish one mosquito from another.

In 1901 the first Nobel Prize for Physiology or Medicine was awarded to Emil von Behring "for his work on serum therapy, especially its application against diphtheria." When considering the next prize, the Medical Faculty of the Karolinska Institute had before them twelve nominations of Ronald Ross from seven different countries, including one by Lord Lister, then president of the Royal Society of London, who nominated him "for his discovery of the development of the malaria parasite in the mosquito." Three of the nominators proposed Grassi jointly with Ross; one, H.F. Bowditsch, a physiologist at Harvard, nominated Bastianelli, Bignami, Grassi, Ross, and Manson all together. The Medical Nobel Committee of the Karolinska Institute met on 25th September, 1902, and decided that fourteen of the candidates who had been proposed should not be considered in that year. That list included Grassi and several other famous men who were to receive the prize later, including Robert Koch, Paul Ehrlich, Alphonse Laveran, Jean Bordet, Camillo Golgi, and Ilja Mecnikov. This left only three candidates. One member of the committee, Professor E. Holmgren, then expressed the view,"Of the three scholars among whom the Committee agree that the choice of prize-winners should be made - namely I. Pavlov, N. Finson and R. Ross - I consider that Ross by reason of his researches into malaria, which are of the greatest practical and theoretical importance, should without reservation be nominated as the most meritorious candidate." Yet, for the 1902 Nobel Prize, the prize committee of the Swedish Academy of Sciences had shown an intention to award the Nobel prize of Physiology or Medicine jointly to both Grassi and Ross in recognition of their discoveries of enormous importance for global public health, and was therefore very embarrassed by the dispute. In response, it appointed as "neutral" arbitrator a scientist of great authority and expertise in the specific field: Robert Koch, who had carried out research on malaria in the spring of 1898 at Grosseto, the main town of the Tuscan Maremma, a malaria region par excellence. On that occasion, he had argued with Grassi, who had disagreed with some of the German microbiologist's analytical methods and deductions. Koch had still not forgotten the ill treatment he had received at the hands of Grassi several years ago during an invited visit to Italy. Now he had got an opportunity to settle a personal score and wielding his tremendous scientific influence, maneuvered to vote against Grassi in the final round. The logical consequence was that Professor Grassi was most unfortunately dropped as Koch's arbitration was in favour of Ross who received the prize in 1902. Finson got the Prize in 1903 for his treatment of lupus vulgaris with ultraviolet light, which was used for more than forty years, and Pavlov in 1904 for his work on the physiology of digestion. Laveran, who should really have received the prize before Ross, was finally awarded it in 1907, after he had been nominated by Ross, as well as several French academicians. Grassi does not seem to have been considered again.

The 'Mosquito World War' which had already begun around early 1890s now reached the climax - a point of no return. Henceforth both Grassi and Ross battled on virtually every issue of priority in several discoveries in malaria research they encountered. Influenced by the widespread British Empire and, in part, Ross's own genius in effecting timely scientific publications and/or exposition of his achievements in established scientific forums, mostly through generosity of the ever helping Dr Manson, in Great Britain, Ross was destined to be more widely recognized and rewarded. Ross was said to be so insolently obsessed with squeezing all the possible honours that he eventually marred his highly affectionate and long-time relationship with Dr. Manson whom Ross had been calling his mentor and teacher. When Manson wished to highlight the malaria research contributions by Grassi, Ross harshly discounted Manson's science and even suggested that Manson hindered rather than helped his malaria work. Thus, the Great Malaria Mosquito War had set in with full-blooded attacks and counterattacks over the primacy of mother of all discoveries, that is an *Anopheles* mosquito obligatorily transmits the malaria parasite from man to man.

Scientists in Italy had a long tradition of working on insect-borne illnesses, including 'intermittent fevers' allegedly resulting from the widespread 'mal-air' emanating swamps, dangerously prolific with mosquito productions. The scientific community in Italy was blessed with a string of geniuses like Grassi, Celli, Golgi, Marchiafava, Guarnieri, Dionisi, Bastianelli and many others. Yet, in spite of all the treasure of data generated through long years of dedicated perseverance, they would best be known to have won in the last lap of the medley by pinpointing irrefutably the role of particular *Anopheles* species (in this case *An. claviger*) in mediating malaria transmission as well as the malaria cycle, in human. In a parallel to what they earlier did following Laveran's discoveries, they repeated history by completing, rounding out and refining Ross's discoveries. And, therefore, even with a massive scientific background they lagged behind a solitary medical scientist in India.

In working out a connection between malaria and mosquitoes, at least two follies were significantly discernible in Italians' endeavours. First, they procrastinated publication of their findings. If only intensity, dedication and volume of work done is considered then Prof. Grassi, the most eminent living zoologist then, would certainly have finished first in the race to pin claims to the discovery of the transmission of malaria by the mosquito. But no, he reserved his well-studied judgement. Rather he busied himself in arguing that the parasite, if it were a parasite, had to have a nucleus and none had yet been found. He did not cease at it and further argued that it had to eat and no one had definitely shown that it did. So, little realizing about Ross's amazing writing prowess and an insatiable hunger for seeing his manuscripts published at the earliest, Grassi 'waited' for the last tittle of evidence which was accomplished only in 1889 after several scientists' researches. Later on, picking up the old threads, the Italians began work in the middle of July, 1898. By that time Ross's proof was complete and partly published. By the end of summer, Ross was able to see his results made public both by circulating his official report to all those working on the subject of 'mosquito-malaria connection', including the Italians, and the lecture that Manson presented in Edinburgh, printed in two reputed journals, *The British Medical Journal* and *The Journal of Tropical Medicine*, in 1898. Secondly, none, even the preeminent Grassi, stuck to the 'mosquito-malaria connection' on sustained and long term basis, and there were many gaps in resuming researches. The kind of unfazed zeal, focused dedication and resolution, and maddening obsession that Ross exhibited in every action, verbal or written, was nowhere seen at any stage of 'mosquito-malaria connection' research by the Italians. Black sheep and scientific charlatans, who existed everywhere looking for suitable opportunities, lost no time to impugn scientific credibility of those toiling hard to save the humanity by unraveling malaria mystery. Consequently, the duel Grassi-Ross war turned more ugly with the introduction of great German bacteriologist, Dr. Robert Koch, who too was tempted to have made the claim as the first to make the 'mosquito-malaria connection' discovery as early as 1883, without putting forth much proof, though! Finally, the rivalry took a triangular tangle, where Grassi was sandwiched between the giant Koch and the flamboyant Ross. This equation had its own very unfortunate and eventful bearing on a later date. Subsequently, when the Nobel Prize jury met to select the winner for the 1902 Prize in Medicine and Physiology, it was Koch, who was himself locked in animus with Grassi and was filled with jealousy ever since his Italy visit and Grassi's hurried publication of *Plasmodium* cycles in *Anopheles* mosquito and human in 1890 without any reference or acknowledgement to Koch's own hypothesis, rallied against the idea of conferment of the medal jointly to Grassi and Ross. The result: Grassi was dropped!

Truly speaking mosquitoes were suspected for long to have connection with certain illnesses, particularly intermittent fevers. Malaria was already known widely and adequately characterized clinically amongst the folks in various parts of the world. It indeed occurred as a special malignity of the environment. Europeans did know through expeditions to South America from 17th century onwards that cinchona bark extraction was able to cure intermittent fevers but they had classified all fevers as one phenomenon. Europeans, whose dominance around the world then was phenomenal, associated malarial sickness with wetlands, hence the name '*paludisme*' [which in French means from swamp; *palus* (Latin) = swamp]. Rome was particularly a malarial inferno owing to Pontine swamps in vicinity, hence the local name 'Roman fever'.

Clearly malarial fevers related to swamps, as unhealthy and notorious fever breeders, along with disturbed lands, forests under denudation and lands under ploughing, particularly the last one, hence the emergence and acceptance of the "telluric" and 'miasmic' theories for centuries. Yet, though traditional epidemiologists stuck to their unhealthy soil (telluric) and malicious air (miasmic) theories, some noted 18th centuiy Italian physicians, like Lancisi, tried to find out etiology of the disease by explaining the fevers were caused due to inhalation of a living contagium or animalcules, called 'serafici' invisible to naked eyes, which multiplied in human body under commensurate conditions. A hypothesis subsequently was advanced by Rasori who noted that different kinds of fevers were caused by different types of species - an extraordinary thought so close to the true interpretation of malaria which would still take several decades until discovery of advanced microscopes good enough to spot micro-organisms. The relationship between swamp-emanated inhaled bad air, comprising whatever toxin of Lancisi and others, with the fever was so obvious that the Italians, possibly in the 1851 century, naturally saw the cause of fever as malaria (bad air), which was later refined as malaria in Great Britain. Subsequently in late 19th century when Laveran, and others, discovered causative parasites, *Plasmodium* species, the word 'malaria' was readily accepted unchanged for the disease itself.

Both the telluric and miasmic explanations circulated for a long period, added by another explanation of mixing of the toxin or poison from the soil into water which resulted in fever after being drunk. This hypothesis changed the course of thinking about malaria which was now classified along with other important gut diseases as cholera and typhoid. This meant that, as typhoid and cholera - which were associated with unhealthy aquatic, land and living environments - had to be taken care of through improved sanitary measures, malaria too - considered as a disease of rural decay, spawn of an ill-kempt, savage and uncultivated land, yielding to good husbandry and civilization - fit the pattern and invited similar countermeasures. This thinking of control of disease through good sanitation is still largely practiced.

So, water was understood beyond doubt the key factor to malaria prevalence. Interestingly, it was discovered inadvertently that malaria distribution virtually overlapped the distribution of anopheline mosquitoes. Why the miasmas were associated with these mosquitoes is due to the fact that most of the malaria-carrying species bite following the night setting in, a pointer to the fears surrounding night miasmas. Another interpretation is the feeble-bodied adult females, in particular, face the danger of desiccation if exposed to the heat of the day, hence they venture out and abound more often when the air is damp than when it is dry which is very much likely during the daytime. While one school of thought of the day prudently tried to build the connection between intermittent fevers and the mosquito prevalence, another school silently advocated the presence of the pathogenic micro-organisms. Unfortunately, the fewer doctors who ventured out to bridge observed associations between mosquitoes and intermittent fevers were actually scientifically greatly impoverished due to lack of demonstrable facts. Some approached the solution in a different way, Dr. John Crawford, an American physician, while working on the 'Yellow Fever Mosquito Theory' opined in 1807 that these are the eggs 'insinuated into our bodies' which caused the malarial fever. He was faced with all round criticism, but little did the world know that he was more nearly right than anyone else would be for ninety years in future. In 1854, Louis Beauperthuy, a French physician, better known as the prototypical author of the mosquito theory, boldly connected mosquitoes with intermittent fever - in warm weather but not in the cold. He was however firm in narrating that mosquitoes injected toxins, picked up from the putrefied vegetation in swamps, into humans. Rasori, an Italian physician, thought mosquitoes might carry the contagium or animalcules. All these early thinkers had no idea or a scrap of evidence of what exactly that imaginative contagium was or how and where it came from, albeit the basic notion that parasites might shuttle between different animal species having been already proposed in 1790 by Peter Christian Abilgaard!

Mosquito has been conceived to cast evils on man for centuries and by 1870-80s some concrete evidences also began to emerge manifesting mosquito's role in transmitting certain diseases. The first important work on the role of insects was done by Dr. Patrick Manson in China during the 1870s when he showed that mosquitoes might be vectors of the disease colloquially known as elephantiasis from the gross swellings, particularly of the legs and scrotum, that it often produces (Harrison, 1978). The hypothesis underlying mosquito's role in transmitting filariae precisely dawned on Manson in 1875 when on a furlough to England. Reasons which he forwarded were absolutely irrefutable: he found out the young microfilariae, which unlike their procreator moved freely in blood stream and were always enclosed in a delicate membrane, could neither feed nor move of their own volition, and therefore must complete their life cycles somewhere else - the mosquito?! But, Manson whose inquisition into the ingenious ways of nature was ebullient and effervescent conjectured by working meticulously and arduously for seven long years on his 'Filariasis Mosquito Theory', all along with his fulltime and highly successful medical practice, and despite excruciating illness due to general gout which he contracted at the age of twenty six and suffered from it throughout his life - often totally crippled for days or weeks. Yet, finally in early 1880s, Manson succeeded in demonstrating that mosquito, *Culex pipiens quinquefasciatus*, was responsible to effect man-to-man transfer of the not so deadly but fiercely dreadful and debilitating lymphatic filariasis worm. This discovery was by all standards a great victory in furthering human knowledge about mosquito's placement in the man-mosquito-disease triangle. However, despite the partly true nature of his discovery (Manson, a surgeon-physician, had made two major mistakes while proclaiming mosquito's intermediate role in disease transmission: (i) First, he thought that these worms completed their development in mosquitoes by five days after feeding - which we today know is more than 10 days - which implied that worms had still not matured, and (ii) Secondly, believing that the mosquito fed only once in her lifetime, he postulated that the parasites infected people only when they consumed drinking water having disintegrating body of an infected mosquito), it bore no adverse significance, considering Manson's phenomenal recognition as the tropical medicine genius and his firm demonstration that mosquitoes performed an essential role in generating human diseases, particularly his guidance to Dr. Ronal Ross about malaria and mosquito - a subject that would soon reveal the mosquito's role in producing explosive outbreaks of the disease, both in terms of human maladies of morbidity and mortality exacted by the *Plasmodium-Anopheles* nexus, as well as the unprecedented show of scientific rivalry by geniuses of the day. The filariasis-mosquito discovery by Manson went largely unopposed.

Taking cue from his 'filariasis-mosquito connection' theory, Manson began to speculate on malaria. Like Laveran, he too was baffled with the encounter of 'exflagellation' phenomenon of the parasite. Following Laveran's assertion that these flagellae were the mature stage of parasite, as well as the speculation that mosquitoes picked them up in that stage for transfer to a new host, Manson, in 1894, began to navigate his thinking the filarae way, though quite faultily. Already, for the benefit of Manson, the emergence of flagella in only half a minute, or more, old drawn blood was highlighted by the Russian Sakharov in 1889 and confirmed by Grassi in 1892. By this time, too, Manson - a warm, patient and avuncular doctor of 50 years of mature yet youthful age - had emerged without question the outstanding expert in tropical diseases in England, and it had become widely known in medical circles that he, next to his remarkable 'mosquito filariasis theory', was seriously embarking to explain the malaria tangle by way of mosquito's role in transmission. However, Manson's newly found filaria-related mosquito theory weighed heavily on him and he drew the conclusion that the red blood cells which enclosed the malaria parasite were comparable to the filarial sheaths. Consequently, he landed with a totally misplaced explanation. Even so, Manson's mosquito theory (first presented primordially in 1894 and later in the fully developed form in lectures of 1896) was a brilliant imaginative construction to pave the way for future research. Ross picked up the evidence where Manson had left!

Almost around the same period, precisely in 1880 in Havana, Dr. Carlos Finlay aired his finding of *Aedes aegypti* (then known as *Stegomyia fasciatia*) transmitting the Yellow Fever disease. He arrived this adrenaline-charged conclusion by pursuing a scientific methodology, otherwise next to impossible at present due to ethical reasons. He made yellow fever patients to be bitten by mosquitoes, only to be let loose to interact in course of time with the uninfected healthy persons. He did succeed partially in effecting yellow fever in some healthy subjects, but failed to realize the minimum time period the infected mosquitoes needed to be fully infective (actually, Finlay only conjectured about the disease based on its jaundice-implicated nature impaling the patient, without any knowledge about the pathogen, a virus, which was not yet known to the science). But, unlike Manson in case of filariasis mosquito, Finlay rightly discovered that mosquitoes bite more than once - a trait that truly pointed out towards mosquito's role as a biological transmitter of the disease. However, as it would normally go with every exciting new discovery, Finlay was not only widely ignored, he was ferociously repudiated for this observation of his - a Cuban, particularly by the then dominating English and German workers, but also by his contemporaries in Latin America. Finlay's yellow fever mosquito theory meandered a long journey through all thick and thin to be finally accepted, almost in the name of another person, a truly famous mosquito-man himself though, the great Dr Walter Reed, when in 1990 a universal acceptance was accorded to this great piece of scientific inquiry. Such was the deep impact of world-wide recognition of the scientific genius of Dr Reed, a U.S. Army officer, that the 'discovery of Yellow Fever mosquito theory' was without any further innuendoes accepted globally. However, unlike malaria where the pathogen, a protozoan parasite belonging to genus *Plasmodium*, was discovered much earlier than its mosquito vector, the world had had to wait for some thirty years before the yellow fever virus could be identified, along with another discovery regarding feral primate populations in the American and African forests harbouring a reservoir of the disease. Luckily to Finlay, his notion on mosquito theory for the yellow fever was propped up by the ultimate discovery of mosquito's prowess as a vector of malaria by Ross and the Italians, amongst which Grassi was most prominent, in 1897.

The three major mosquito-related researches in connection with demystifying transmission cycles of some of the most deadly and debilitating tropical diseases included well-placed and rather globally important 'Mosquito Yellow Fever Theory', 'Mosquito Filariasis Theory' and 'Mosquito Malaria Theory' - all during the fourth quarter of 18th century. Of these three milestones, the yellow fever and filariasis connections to mosquito went rather peaceful or unchallenged, though low-voiced disagreements or oppositions could always be audible. Both these mosquito-borne revelations were made much earlier than the one for malaria or, more importantly than anything else, the event of establishment of the highest prize money in the name of Nobel Prize in 1896, to be commenced for awarding from the year 2001. Malaria thus asserted some special claim, as already was paralleled - with still greater attention due to its prowess to stay on year after year - with the population-decimating deadly plague, a reputation no other tropical disease was bestowed with! Enemies not in clear vision then included *Plasmodium* parasites and the *Anopheles* vector mosquito species, and, true to the demand of Nature, they were to be found out by none other than the two military men, Doctors Alphonse Laveran of French Army and Ronald Ross of British Indian Medical Service, who made basic discoveries that unlocked the secrets of malaria. Both of these scientists had characteristic opposite basic human qualities (Harrison, 1978). Dr. Ronald Ross, a complicated and chronically maladjusted man, was nevertheless filled with ceaseless energy, endurance, positive thinking and, above all, the desire to do something extraordinary - something that nobody had earlier done and which would single him out amongst the world's greatest men of honour. He took life as a struggle and a challenge, and himself as a soldier eternally pitched against the people and forces who sought, from stupidity or maliciousness, to frustrate him and hold back the course of human progress. Ross was exaggeratingly ambitious for progress and would not accept anyone as hindrance in his way. Dr. Alphonse Laveran was by nature a sober and noble human having a special penchant for scientific research and professional progress.

Initially it all started as a pure international competition, with the noted German pathologists Dr. Heinrich Meckel, in 1847 and Dr. T. Frerichs, in 1858, along with Dr. Rudolf Virchow, having described and characterized the black particles in the blood smears of malaria patients of being formed of degraded haemoglobin from the affected red blood corpuscles, respectively, and many a French doctor like F.C. Mallot, Leon Colin, P.J. Keiner and Acheille Kelsch, particularly the last one, who had enormous experience in malaria pathology, showing presence of innumerable tiny black particles, especially conglomerated in spleen, liver and, sometimes, brain to give the affected organ a tinge of slate grey.

Many explanations were floored as to the origin of these malaria specific black particles, and the science of understanding moved ahead without much hassles. Dr. Kelsch, who worked in Algeria in early 1870s, apparently missed the discovery of the first malarial parasite viewed in the human, which was some six years later done by Laveran in Algeria only. Professor Kelsch, true to the nature of his profession of a pathologist, looked at everything malarial with a classical pathologist's eye only, even having probably routinely killed the contents of the freshly drawn blood from the malaria patients - in which there existed every likelihood of finding what Laveran had subsequently discovered, the 'crescents'. Though, Kelsch had passed by the brink of discovery, he nevertheless had demonstrated that the black pigments emerged with the onset of fever. He obviously did not follow the question of uniqueness of black pigment in malaria, and focused on other important lines of malaria research pertaining to the disease's impacts on red and white blood cells. Thus, the central theme was left unattended and unraveled, for the man who had a keen eye for observation and a determination to find answers to most baffling queries - Charles Louis Alphonse Laveran.

Laveran, sent to Algeria in 1878 as an official necessity and certainly not on any mission to research on malaria, was nonetheless forced to go into studying malaria as military bosses were desperately keen to cease the outbreaks (of malaria) that constantly wrecked their soldiers there and exacted astronomical costs in replacing dead and the sick as well as in boosting up their soldiers' morale. He was familiar with the work of Kelsch and others and soon understood the value of studying alive the contents of the freshly drawn blood to which he had access in plenty. He met with success rather early on 5th November, 1880 when he clearly saw 'spheres' (the exflagellating male gametes). For the next four years he worked to know more about the malaria parasite which he described *in extenso* in a book titled '*La Nature parasitaire des accidents de l'impaludisme*', published in 1881, followed by two more elaborative books in 1884 and 1898, respectively. When he reported his findings to the Academy of Medicine in Paris he did not forget to underline his claim to have been the first to discover the malaria parasite. It was indeed the first glimpse of one form of the malarial parasite, *Plasmodium malariae*, a protozoan - still not known by that name. Ironically, even such a spectacular discovery had to go unattended in appreciation. There were contemporary scientists who stooped low to belittle the significance of Laveran's grand discovery by giving one or another bizarre explanation. Others, more vocal, including the celebrated haematologist Dr. William Osler, Baltimore (USA), totally refused to believe his observations and an event such as the 'exflagellation' in blood. Disagreement with Laveran's single-celled protozoan (which he referred to as the germ) led to divert the attention of several scientists of the day to consider an alternative possibility - the bacterium. The bacillus theory of *Bacillus malariae* prevalent in Italy's infamous malarious swamps, raised by two Italians, was as soon sidelined as it arose in spite of undercurrent of support from a newly emerging highly promising discipline of bacteriology directly under the guidance of as immensely famous bacteriologists as Louis Pasteur and Robert Koch. However, this dethronement of malarial bacillus hypothesis did not quite help ready acceptance of Laveran's germ theory but, on the contrary, Laveran had to take upon himself the responsibility to convince in person by practical demonstration of 'exflagellation' to the fellow scientists both in Algeria (Dr. Richard) and in Italy (Drs. Ettore Marchiafava and Angelo Celli), but largely in vain, since both the Italian experts had their opinion reserved in favour of *Bacillus malariae*, thought as the most likely agent of malaria. Not only this, Marchiafava and Celli mischievously hid from him their own research outcome in next two years conforming to Laveran's findings, which they published without any reference to Laveran's remarkable original work with a view point of making their claim as first for the discovery. Laveran was immensely exasperated by this fraudulent act of Italians who, in turn, believed what they experienced in fresh blood preparations was not seen by Laveran and was certainly different. In an attempt to put their stake for original claim of the discovery they named their form as *Plasmodium malariae.* Dr. Camillo Golgi, whose primary interest was the nervous system, confirmed their findings, with some startling advancement in knowledge. With this the initial coolness of trust in comradeship and scientific fraternity seemed to have volatilized, and the first seeds of an embittered future world war of priorities and primacy had been sown sans healthy international competition even when new human malaria parasite species continued to be discovered with a large number of excellent scientific inputs by Golgi and others.

Laveran, as said earlier, was a never-say-die type of scientist and his consistently good relation with scientists worldwide paid dividends before long. Starting in 1885, Laveran's discovery soon spread to America, Great Britain and India, and by the end of 1890 it was said that acceptance had been universally granted. The technique with a lot of malaria-related knowledge would in time passed on to Dr. Patrick Manson in Great Britain who, in return, descended the wealth of facts to his protège, Dr. Ronald Ross. Both Grassi and Ross would later take their own time to accept or reject these wonderful discoveries, but never to sit quite even a bit moment when it came to their own research findings. The stage for gruesome "First Malaria-Mosquito World War" had just been set in.

Dr. Ronald Ross, who had been ushered into the medical profession against his intention to become a writer or a musician, did not even imagine ever to pursue research in his early life in British Indian Medical Service, leave aside the exemplary Nobel Prize winning mosquito-malaria research which of course he did beating everybody else with his unmatched discipline and focused attention. The transformation was not subtle. In fact, Ross seemed to have never come to terms with the extreme tropical climate of India, rife with blood-sucking mosquitoes everywhere almost round the year. It was after all his father's plan, not his, to serve in the British Medical Service in India where the senior Ross was lucky to have risen himself to the rank of a full General! Ronald Ross, a good son, obeyed his father to this extent, but felt eternally distant from medicine and research, and quite closer to matters deeply imprinted on his mind - the music, poem and novel writing (he did succeed in publishing a romance novel, *The Spirit in the Storm*, in London in 1896) and, of course, mathematics. In his initial days Ross had felt tormented by both the dusty heat of Indian plains (he once even got a brief posting to Nasirabad close to the eastern edge of the Thar Desert in the erstwhile Rajaputana Estate, now Rajasthan) and the callous behaviour of his seniors in sending him quite frequently on postings to less cozy and recreational, and monetarily least attractive places, still mostly as a forced bachelor having had to leave the family behind either in the native Great Britain or the previous posting. Consequently, he silently confronted his military bosses on one or another pretext, and, out of agony and exasperation, even threatened to quit the military job and return to England. Frustration enveloped Ross - *hook, line and sinker*! He had plenty of time to golf away and compose music, but the impact of continued humiliation in service kept constantly blasting off his ever alert mind to do something extraordinary that nobody in the world had earlier done and doing what he could rise to the top honours of the world. He was eternally restless but did not know where to go, or navigate his life to reach his objective. Greatly lost in the labyrinthine medical disciplines and researches, he decided to coolly think meticulously within the precincts of his medical assignments and carefully narrowed down to malaria research which indeed was the hottest topic of the day! Ross immediately jumped with a burning desire in his eyes. He thought he had got the objective, but still did not know how to strike the bull's eye! The problem was, however, that Dr Ross, with almost total ignorance of the literature, knew very little about malaria, had never seen a parasite, did not know the necessary techniques to examine malaria parasites (despite the fact that Laveran and, subsequently, the Italians and Americans have demonstrated microscopically the presence of various stages of *Plasmodium* species throughout the second half of the 1880s) and could not at all distinguish one mosquito from another.

Overcame he did yet most of these one by one as the inexorable laws of time and destiny had in store for him something the whole humanity will be later proud of. As if providentially designed to reach his objective, he first met the great Dr. Patrick Manson under whose careful tutelage Ross rose to enormous heights of success, winning finally the race to pinpoint *Anopheles* mosquito's role in malaria transmission in 1897 and, more rewardingly for Ross, the Nobel Prize in 2002 (actually handed out in 2004 due to long drawn dispute). His discoveries of the role of *Anopheles* mosquito (which he, ignorant of any scientific nomenclature, colloquially referred to as 'dapple wings') in human malaria transmission and the full malaria parasite life cycle in an avian model are a classical saga of exemplary determination, perseverance, courage, patience and ambitiously focused application.

At a time when Ronald Ross was groping in dark and ceaselessly endeavouring to grapple with the necessary microbiological and microscopic techniques, the Italians, particularly the genius of Giovanni Battista Grassi and his team of as celebrated scientists as Drs Marchiafava, Celli and Golgi, were already at the problem of solving the malaria mystery through implication of mosquito. It appears that both Ross and Grassi knew about each other's moves in demystifying the hidden secrets of malaria cycle, but none really showed any signs of this awareness. Ross, true to his military training was meticulous in keeping his day-to-day account of mosquito-malaria experiments as well as bringing the findings to the notice of all through Manson in England and quick dispatches to the scientific journals of international fame. Thus, not only his phenomenal express reporting of the pigmented cells in two mosquitoes fed on malaria patient (Ross, 1897), even publication of his proof of mosquito's involvement in an avian model did not take long to get public within a few months since revelation of pigmented bodies in *Anopheles* mosquito (Ross, 1898a,b). He carefully planned his bracing the tape first by every possible, logical means in the international race for staking the claim for the unique discovery that coincided knowingly or unknowingly with the news of institution of the Nobel Prizes, including one in Medicine and Physiology, following the death of their willer, the dynamite king Alfred Nobel, in 1896, which, if won, meant unimaginably colossal prospects ahead both monetarily and from honours point-of-view.

Around the same time, and possibly a bit earlier to Ross's revelation of August 20, 1897, Professor Grassi, who had been working relentlessly on the mosquito-malaria connection for several months, then immediately unaware of Ross's discovery, true to his reputation and genius, patiently waited for the last iota of truth to emerge to foolproof his own 'first' discovery of the same matter as Ross's. Grassi's first publication from his observations taken since mid-July, 1898 through summer 1899 came to see the Sun in October, 1899 and then in December, 1899, wherein he announced, referring to his associate colleagues

(Celli, Bignami, Dionisi and Bastianelli), having arrived at a definitive solution of the question of mosquito-malaria riddle and that the victory was completely theirs. The rush to publish the results and air the news was unusual and inclined to beat Robert Koch who too had been advocating his own mosquito-malaria theory since as early as 1883. Ironically, it was Ross, and not Koch, who was hit. Ross felt treacherously ignored by the Italians for his 'first and original' discovery. He questioned the originality of Grassi's research, insisting that his indication that a "grey mosquito with dappled wings" was responsible for the transmission had guided Grassi in the identification of the vector. He even accused Grassi of fraud, on the basis of incorrect dating of the notes presented to the Lincei Academy, which instead had been precisely certified by the date of presentation in the Academy's public session. Grassi reacted harshly to these accusations, which he felt offended his honour as a scientist.

Italians' superior techniques and work infrastructure were always highly revered in scientific circles. Even Ross was full of praise for both Grassi's and Bignami's ingenuity in medical research and in identifying the vector, and confirming the mosquito cycle of the *Plasmodium*. In return, Grassi returned courtesies full of applause to Ross for his efforts. These pleasantries between Grassi and Ross were only ephemeral and all the sense of mutual respectability effervesced with the initial news aired about Ross's having brought first his claim for the primacy of the discovery of *Anopheles* mosquito being invariably responsible to transmit malaria. The competition readily reached its lowest and ugliest point with the war for credit of priority and primacy of the discovery. Ross charged Grassi's work of several years as childish which he could have accomplished in a few weeks. Grassi, who was rightly considered by many to be the preeminent medical zoologist of his time, shot back calling Ross as an amateur and an interloper. Ross was stung to an instant wrath. He was a person once launched in anger was never to pull back.

Dr. Ronald Ross and Professor Giovanni Battista Grassi envied each other for the rest of their lives, broaching over priority in various different mosquito-malaria related researches. Ross, feeling eternally insecure for his place on various discoveries, was more vociferous and hostile, having spared no opportunity to attack Grassi and inflating his own achievements. Ross engaged in a lifelong polemic to quash all Italian pretensions and reassert with ever-diminishing qualifications that he and only he had demonstrated the transmission of malaria by mosquitoes. Grassi, who in course of the continuing controversy on primacy, lost association of two of his best colleagues Bignami and Bastianelli – the chief architects of Grassi's mosquito-malaria work, due mostly to his growing arrogant and egoistical contentions, battled Ross ceaselessly with matching intensity. Ross's obsession with garnering every possible laurel was so intense that he did not even hesitate to hurl insinuations on his mentor - Dr Patrick Manson, his science and scientific integrity, just because Manson, truly a generous, impartial and honest grand man of tropical disease sciences, indicated to credit Grassi for his outstanding contributions in malaria research. Ross, in spite of having lost association of many friends in course of the never-ending battle with Grassi, did have many others to support his cause and value. One such well-wisher was Dr. Robert Koch. Haunted by Grassi's ignominious treatment to his Italian visit in late 1890s and later the surreptitious correspondence, Koch had turned into an eternally implacable foe of Grassi. Koch's persistently vehement attacks on Grassi's scientific integrity, climaxing into his violent opposition for the Italian genius against a consideration for awarding the 1902 Nobel Prize jointly to both Grassi and Ross, in all likelihood cost the former the fortune. Ross won the Nobel Prize, besides the prestigious Knighthood a few years later. *Still, if a logical conclusion is drawn it would become crystal clear that though Ross was the first to show that malarial parasite was transmitted by the bite of infected mosquitoes, in his case the avian Plasmodium relictum, nevertheless, it was absolutely certainly inevitably Grassi's work which revealed that human malarial parasites were carried only by female Anopheles mosquitoes. He identified the mosquito species correctly, in his case Anopheles claviger. By today's standard, they are likely to have shared the Nobel Prize.*

For whatever happened between these two geniuses the humanity was greatly benefited from their tireless inquiries in the mystique mosquito-malaria relationship and what benefits we reap today in saving precious lives to the savage attacks of the insidious malarial disease is due entirely to their, and other similarly devoted scientists', tireless and focused hard work.

"Problems cannot be solved at the same level of awareness that created them."

● ● ●

PART IV: MALARIA TO KALA-AZAR

17

In Assam on Non-malaria Mission: Ross Emerging a Complete Scientist – *Bad Times have a Scientific Value*

"Success is a science;
if you have the conditions, you get the result."
– (Albert Einstein)

During Ross' tenure in Calcutta for investigating the matters related to the transmission cycle of malaria parasites, the Indian Medical Service again showed lack of support towards his commitment to malaria work and instead assigned him a new duty to investigate kala azar in Assam. Ross was then not in a mood to step aside of his malaria research and demanded an extension for 6 months to work on the malarial parasite transmission cycle. His appeal did not please his bosses who not only rejected his appeal but instructed instead to move expressly without availing obligatory joining time to Assam. After this it was no longer possible to postpone the evil hour without the charges of dereliction of duty, and he was obliged to set out on the long journey to Assam. Ross's consternation was comprehensible but he could in no way muster in enough courage to oppose the orders. On 13th September 1898 he went to southern Assam in (northeastern India) to study an epidemic of visceral leishmaniasis, also known as Kala-azar.

Dr Ross (Fig. 17.1) was novice to Kala-azar or Visceral leishmaniasis for he has encountered several patients during his Darjeeling sojourn. Consequently he obtained permission to investigate this disorder first. Fixing his headquarters at Kurseong in the hills on the road to Darjeeling, Ross made numerous visits to this locality, but was met with failure entirely. The plague-scare, though waning, was still present; and difficulties of transport impeded the work. On the 25th August I arrived at Naxalbari, an intensely malarious plantation and village on the plain beyond the foot of the hills, and found swarms of small and large dappled-winged mosquitoes (probably *Anopheles listoni* and *An. rossi* ?). He examined some dozens of these mosquitoes caught in the houses of infected persons, both for the pigmented cells and the thread-like bodies, but in vain! Nearly all his time was actually taken up in pathological enquiries on Kala-dukh (Black sickness).

Fig. 17.1. Major Ronald Ross, photographed in his study, with microscope.

After arriving Nowgong, the center of the epidemic of Kala-azar in Assam, he was invited to work by Dr Graham Col Ville Ramsay, the second Medical Officer of the Labac Tea Estate Hospital (where his microscope and medical tools are still preserved, and his sketches of mosquitoes are on display) (Fig. 17.2; Dr Soumya Swaminathan, Director General, Indian Council of Medical Research and Secretary, Department of Health Research, Govt. of India Soumya Swaminathan visited his laboratory 6th Oct., 2017 and suggested to develop into a museum). He learned that the disease Kala-azar is a difficult pathological problem, completely different than that of malaria, albeit being caused by a protozoan. He certainly feared hugely of getting involved in preparing long pathological reports for the government, although soon his passion for the disease surfaced and he began to find it as interesting as his first love, malaria. Apparently there were many challenges associated with the disease paralleled to those of malaria; the disease was first diagnosed as ankylostomiasis and later even a virulent form of malaria. Ross was asked to find out if the disease was communicable? It was a tough task for judging as Ross himself personally believed the disease malaria.

A mélange of kala-azar and malaria cases occurred. In advanced stage a Kala-azar case could be easily misdiagnosed as fatal form of malarial cachexia by the local practitioners. As his task was indeed a complex one he attacked the problem by examining the blood of all the cases, both of malaria and of kala-azar. His results showed that while the parasites were easily found in the early cases, they became scarcer as the disease advanced until, in the old typical cases of malarial cachexia and kala-azar neither parasites nor pigment were to be found, even in blood taken from the spleen. He repeated examination of the blood of all the cases which he could procure in the town. Long hours of microscopic work caused much strain on his eyesight. He believed that the Kala-azar parasite, *Leishmania donovani* discovered by Leishman and Donovan (Ross coined the nomenclature *Leishmania donovani* in 1903, in the honour of Dr Leishman and Dr Donovan, to end a long-standing puzzle), was transmitted by a dapple-winged mosquitoes mosquito (possibly *Anopheles rossi*). Since Ross was already exposed to recent historic malariological explorations, he seemed naturally biased toward malaria and began his work with a predisposed inclination of mind. Consequently he collected and dissected scores of dapple-winged mosquitoes (possibly *Anopheles rossi*) but without any success. Moreover, he had absolutely no idea about sandflies – the real culprit in this case of disease. His disappointment was considerable. It is now known that Kala-azar is transmitted by sandflies. On 11th October he submitted a short report on bird malaria due to mosquito bites; this report did not see the sun until several months, however, the results of this study were already made public by Manson in England. At the conclusion of his work on Kala-azar he returned, now utterly exhausted, to Calcutta.

Fig. 17.2. Ronald Ross worked here in Silchar; his chair and microscope intact, as well as drawings of mosquitoes.

Arriving at Calcutta on the 19th November Ross set to work to pick up the threads of his work on *Proteosoma*, to obtain cases of human malaria and to write his report on Kala-azar - this being a tedious business requiring a full discussion of many intricate details. Things had however changed calamitously for him; he was not given any assistants, did not have a room of laboratory, and moreover his superiors were reluctant to acknowledge his achievements. Worst one Dr. Daniels was sent to assist him for nearly two months, although Ross never fully agreed that his job was only to assist and not to enquire into the correctness of statements in his research work and the reports. He was nurturing a feeling that he had done enough in India and possibly it was the right time to go to wife and children in England. He did not yet plan to find an alternative vocation in England but was more than sure that Dr Manson, his father-like mentor, guide and teacher would certainly be helpful in getting him a commensurate job. In India during his nearly two decades' stay he had met with a large number of officers, scientists, assistants and local natives, and cherished beautiful memories on many accounts. He was particularly grateful to Drs. Smyth, the only person, he thought, had constantly taken sufficient interest in his proceedings even at his parasite preparations since 1895. Those amongst several who admired his research acumen included Maynard, Dyson, and Cooke. He felt exalted meeting with Dr. Rivenburg of the American Mission, who, hearing of his work, arrived Calcutta all the way from a distant part of Assam with his wife and children at his own expense, to assist him. More than anybody else Ross was particularly delighted to hear from Manson that his work was taken up by Koch and the Italians, his papers had been published and copies of his *Proteosoma* Report had been sent to many persons interested in malaria. On the 8th November Manson wrote again informing me that he had just dispatched some of my preparations to Rome, namely to Bignami and Charles.

As Ross drew closer his years' special duty in Calcutta, he failed to secure a definite assurance from his chief that he could be retained on the same duty for an extended period. Moreover, he figured out, that if he were then returned to Secunderabad, his original posting, he could not become able later to pay for his passage to England. He, therefore, decided to leave India forthwith and to return to England, trusting to fortune to give him an opportunity for finishing the investigation in a manner which he thought appropriate. He left India in February 1899.

• • •

PART V: BIDDING ADIEU TO INDIA

18

'Farewell India', Dr Ross Bidding Adieu to Return England – *Twice visited the Motherland to pay Gratitude for Winning him the Nobel Prize and much more* (1899-1932)

In the year 1899, Ross found it increasingly difficult to continue his stay in India and make further progress in his research and wanted to leave India for good. His health had by then suffered greatly from the continuous exertion made under very trying circumstances, and he felt scarcely able to undertake any further new posting, with family living aloof in England. The labour, the disappointments, even the successes, of the long and anxious investigations of a single subject had been too much for him. He fixed his sea journey and communicated to Dr Manson by a letter dated 15th February, 1899. He wrote, "We start on the 22nd & hope to arrive on about 20th March - all the way by sea", although in a quick successive letter dated 24th February from the City of Corinth in the Hoogly, Ross updated Manson about the delay of his arrival to about 22nd of March. The later letter drives much significance because in the same communication Ross also congratulated himself on the completion of the task imposed to him by Dr Manson, indicating about his rather recently cultivated obsession with self-aggrandization, on one hand, and sense of great insecurity about the primacy and priority of his research and discoveries – possibly for any yet undecided fortunes in future, if and when determined –, in India (Bynum and Overy, 1998).

On Ross's return to England, it took some time before any commensurate job had come his way. Through sheer benevolence of Dr Patrick Manson the Tropical School in Liverpool offered him the position of a Lecturer. The Liverpool and London Schools of Tropical Medicine were founded in 1899 on the initiatives of the Secretary of State for the Colonies, Joseph Chamberlain, and of Patrick Manson, who was Medical Adviser to the Colonial Office from 1897 to 1912. The Liverpool School depended for funds mainly on private sources and was thus less stable financially than the largely government-supported London School. Ross became a lecturer and later a professor in the Liverpool School (Nye and Gibson, 1897). Ross gave an inaugural lecture which was published as 'The possibility of extirpating malaria from certain localities by a new method', BMJ (1st July,1899). Ronald Ross also brought out a paper 'An outbreak of fever attributed to mosquitos', BMJ on 22nd July,1899. He also sent a brief paper to Nature 'Life-history of the parasites of malaria' which was published on August 3, 1899. Meanwhile Ross was vexed with the terminologies that Grassi used and argued that as a rule the man who discovers the life history of a single member of a group claims priority and since the malaria problem was a purely zoological problem, it is amenable to zoological laws of priority to which Manson responded saying that he liked Ross' terminology and classification and will adopt as he deems fit although zoological laws make it necessary to stick to Grassi's names no matter how ugly they might seem (Bynum and Overy,1998).

Soon after Ross' joining, he wanted to carry out a malaria expedition to Sierra Leone before the Italians and proposed for the same to the committee for which he got only 350 Euros as initial funding but he had to put off his journey to Brussels, a probable reason being that Ross heard no support for his expedition from the Belgium Government. Ross was determined to go to Sierra Leone with as little funding as he got and arranged to go out on the 5th August and return the 15th October having almost six weeks in hand (Bynum and Overy,1998). Later, The Council of the British Medical Association voted the sum of 100 Euros towards the expenses of the expedition, at its meeting on 12 July, 1899 as reported in BMJ (15 July, 1899). During this time 'Surgeon-General Robert Harvey visited the School in Liverpool and suggested to Ross that he return to India to a permanent post and salary' (Ross, 1923). Amidst the expedition being set forth Ross was approached to open the discussion at the sixty seventh annual meeting of the B.M.A. in Portsmouth but he couldn't attend it because he was off to Sierra Leone. Ross gave an account about his expedition, 'The malaria expedition to Sierra Leone, 9th September, 1899, in BMJ which he mentioned in a letter dated 20th August,1899 from Sierra Leone (Bynum and Overy,1998). In the month of December1899, Manson's visit to Italy proved to be fatal to the relationship between him and Ross. Manson was one of the party of doctors and the reports from The Times called Ross an engineer minimising his role in favour of the Italian contribution. While Manson failed to protest to the newspaper, Ross did it himself.

The beginning of the year 1900 saw Ross presenting a lecture 'Malaria and mosquitoes' to the Royal Institution on 2nd March on the invitation of Sir James Crichton-Browne which was published in Nature (29th March, 1900, pp. 522-527), and also by the Institution and Liverpool School. The feud with Grassi increased when Sir Patrick Manson in a letter posted on 8th June, 1900 exclaimed the horror of robbery attributed to Grassi as Grassi claimed to have discovered the filaria in the mosquito proboscis which apparently was found by Ross. A second havoc in Ronald Ross' life came when Leonard Rogers made a report 'Abstract of a paper on the relationship of drinking water; - water logging and distribution of anopheles mosquitoes, respectively to the prevalence of malaria north of Calcutta', in the Indian Medical Gazette on September 1900. Here Rogers although recognised Ross' work, yet argued that proper filtered water supply had a stronger correlation with spleen rate than the mere presence of *Anopheles.* In response to which Ross commented that Roger had no brains nor reliability in a letter posted from Llandulas, North Wales on 9th August. However, Ross decided to give a fitting reply to Roger by publishing 'Captain Rogers' recent investigation on malaria' in the Indian Medical Gazette in December 1900, pp. 461-462, criticising Rogers assumptions that spleen rates were accurate measures of active malaria (Bynum and Overy, 1998).

On the January of 1901, Ross went back to Thomas-Yates Laboratory, University college, Liverpool and received a welcoming honour as Manson proposed Ross for Fellowship of the Royal Society. He was elected on 6th June 1901, at the first time of asking (Election paper, RS) although Manson admitted that he was afraid because being so young a member he might not be the right man to propose Ross's name but others thought otherwise. Ross was later elected FRS on 7th June,1901. In June, 1901 Ross again departed back to Sierra Leone securing a donation of 1000 Euros from James Coats, Jr., to fund the expedition and returned from Sierra Leone and Lagos in early September to the news of his wife Rosa Ross being pregnant with their fourth and final child, Charles Claye Ross who was born on October 13, 1901 (Bynum and Overy,1998).

A year later in 1902, Ross's dissatisfaction with the employment conditions, resigned his post at Liverpool and accepted an appointment to head a new department of animal parasitology at the Jenner Institute (eventually Lister Institute) but Ross ended up resigning his post from Jenner Institute too on June 1,1902 having found the conditions of his appointment unsatisfactory. The year 1902 proved to be a golden one for Ross. He was awarded the Companion of the British Empire in Edward VII's coronation Honours List (Bynum and Overy,1998). The biggest acclaim Ross received was him being awarded the Nobel Prize for Medicine on 10th December, 1902. A notable event during this period was the cold war between Ross and Manson which infuriated from the time Manson went to Italy and it aggravated in a way where Manson received the news of Ross being awarded the Nobel Prize from a newspaper, which was quite unlikely, in spite of Ross receiving the news on the 3rd November, 1902 (Nye and Gibson,1997). From then on a sense of hostility began raging between Grassi and Ross for Grassi had dedicated a paper to Sir Patrick Manson in the year 1903. But Ross felt jealous of him and that got the better of him. Ross showed concerned about Grassi not seeking consent or permission from Manson before dedicating his second paper to Manson because that could otherwise send a false message to the people, of the budding closeness between Manson and Grassi. Manson discarded Ross' argument and asked him to drop the matter immediately and leave Grassi alone and at the least Manson would not want to be a part of a useless squabble and instead suggested Ross to focus the work in hand (Bynum and Overy,1998).

However, Ross wrote his Nobel Prize essay 'Researches on Malaria being the Nobel Medical Prize Lecture for 1902' (Ross, 1904).

During those pink days of his life when awards, laurels, distinctions and recognitions were pouring on him in abundance, it seems so the air of a bit of haughtiness went to his head and he began to deviate from his hard earned habits of resilience, obedience and camaraderie. At this juncture another element appeared on his persona and that persisted for the rest of Ross's life; a preoccupation with money and the rewards of research. So, in 1906, Ross published an unsigned proposal for a British Nobel Prize, arguing that the British should not have to look to Sweden for reward, that fame did not educate a researcher's children or provide him a pension, and that the nation should not use a man's research for nothing; the proposed Prize would pay men who advanced medical science but received no monetary recompense for the work. Ross's plan for a British Nobel Prize failed to stir the medical community. He had, however, mentioned in his paper that in the early nineteenth century Parliament had voted Jenner £30,000 for developing vaccination. This type of reward seemed to appeal to Ross, for he wrote to a friend that "It occurred to me that I had better go to my lawyer and see what he suggested. After I told him about Jenner getting £30,000 from Parliament, he at once said that he thought a case might lie, if not on legal, yet on strong moral grounds." His lawyer's favourable reaction may have deflected Ross from further advocating the British Nobel Prize, and after a lapse of seven years Ross did, in fact, petition Parliament (Chernin, 1988; also vide a footnote there re: 'A British Nobel Prize', Brit. med. J., 1906, ii: 1667-1668. [Unsigned, but Ross privately identified himself as the author in a letter to Sir Alfred Keogh, 3 December 1906, RA 36/007]. He was not greedy, though, nor he was asking for any rewards, but he certainly valued his researches and did not imagine to serve for nothing in return because he thought a work must have a value.

Although Ross had strived to not return to India ever again, he came bersek to India first for receiving the Barclay Bronze Medal on 20th May, 1903 (given by the Asiatic Society of Bengal) and the second time in the year 1909 to attend the Bombay Medical Congress from the 22nd-25th of February to speak on the practice of malarial prevention (Memoirs, 504-5, & Bynum,1994). Ross had also proposed the establishment of a malaria bureau, based at the Liverpool School (Nye & Gibson, 1997). This even tells us about the respect that Ross received but he had no love for the country back and instead looked down upon its people and called it names.

Ross had also visited Mauritius in 1907 to advise on malaria. He believed that malaria had been introduced into the island in 1866, via imported *Anopheles* species. Ross was made Knight Commander of the Order of the Barth in the coronation Honours of George V (Bynum & Overy, 1998). Ross was made Director-in-Chief of the Ross Institute and Hospital for Tropical Diseases from 1926 until his death in 1932 (Manish Pradhan, 2009).

Ross was a polymath and he received several honours, awards and wrote a number of papers during this time (see Annex). The later stages of Ross's life was spent travelling in all parts of the country as means of educating the mass of the measures that could be taken to prevent malaria. Ross went to countries like Greece, Cyprus, West Indies, Mauritius, Suez Canal zone. These places were horribly affected by the World War between 1914-1918 (Manish Pradhan, 2009). Ross was dedicated and committed to his work and that manifests itself in the subtle contexts where Ross takes leaps to help the affected and even after the path breaking discovery of the transmission dynamics of malarial parasite, he continues to work with the same spirit to come off with other discoveries.

Ross was the man for all seasons. Even after Ross found the parasite that could be transmitted via mosquitoes, he didn't just put an end to his childhood passion of poetry and art. Ross wrote versus like "The Setting Sun", "The Triump" and even paid his homage to Shakespeare by writing a play called "Shakespeare". The World War had a deep impact on the mind of Ross and he composed "Versus on the War". This leads us to the way of integrating Science with Literature, a path shown by the polymath years ago. Ross kept on working and committed himself to serve his people. Ross also brought out and published a number of mathematical papers such as "The methods of solving algebraic equations", "Operative algebra" and so on after receiving the Nobel Prize delineating a sketch of himself as a person whose resolved and aspired to keep innovating and working on novel ideas to bring forth a change and create a landmark.

The grass was not always green for Ross. He had to face days that were challenging. Ross remained away from his family for most times in his life because of transfers and also working conditions. He could never fully give his time either to his wife Rosa Ross nor to his children but he always made it up to them and so he also brought his wife and children to Bangalore right after his posting. Ross kept writing for the Indian Medical Journal during this time (Bynum & Overy,1998). Rosa Ross passed away in the year, 1931. Grief-struck Ross breathed his last on 16th September 1932. In his passing away the world, but in particular India, lost one of the greatest scientists ever. The world will remain indebted to the scientific explorations of this inexhaustible genius, Dr Sir Ronald Ross, whose discoveries in mosquito-malaria science have helped to save millions of deaths and infirmities annually.

हम पंछी उन्मुक्त गगन के पिंजरबद्ध न गा पाएँगे,
कनक-तीलियों से टकराकर पुलकित पंख टूट जाएँगे।

(*Source:* Sivamangal Singh 'Suman')

• • •

PART VI: BETTER HALF AND CHILDREN

19. Rosa Bloxam: an apostle of quiet wisdom and strength - ***there is always a woman behind a man's success***

19

Rosa Bloxam: An Apostle of Quiet Wisdom and Strength – *There is always a Woman behind a Man's success*

"Men are a symbol of strength and leadership but at the end of the day, they are also human. They need someone to ask for advice, someone to speak their mind to and someone that is always there for them even when they make a mistake – a woman".

Success is defined as the accomplishment of an aim or purpose. Success (great jobs and businesses) comes with a lot of stress, tough decisions to make and very frustrating times. Successful men are usually ready to face challenges and tackle them head-on with positive energy, the source of which largely originates from their homes. The key player in the home for this man is a woman with strong character that can stand with her husband in prayer during these times, and give him words of encouragement and be the first to believe in him and boost his ego before anyone else does. According to a study men become more productive after marriage; they earn between ten and forty percent more than single men with similar education and job histories. Encouragement and support from a spouse play a key role. Women and men are different when it comes to levels of patience and understanding and when a man and woman form a relationship where they support each other, a man gets a different perspective of things that he otherwise wouldn't have had. And this is key when it comes to fitting into this dynamic world.

'Behind every great man is a great woman'

Acknowledge or not we all know very well in our heart that literally behind every great man there is a woman often remaining *incognito* and being overshadowed by his reputation and defined solely as his wife, mother, sister or daughter. These women – the great women behind great men, some better known than others – have stories of their own, as invaluable partners and companions, sometimes rising to even greater heights of popularity and significance than their male counterparts whose names resonate more loudly through history. At the same time, the contribution of women scientists in science and technology, who – in blatant contrast to men, almost always have to part their times between managing homes – *hook, line and sinker*, on one hand, and the fulltime work in laboratories – with complete dedication, on the other, and still made significant advances to the progress of diverse disciplines of malaria research, cannot be disregarded. There have been numerous gifted and far-famed women scientists in history of malaria who made crucial discoveries and inventions in the science of malariology globally, such as, to mention symbolically a few only, Dr Tu Youyou, (born December 30, 1930; of qinghaosu or artemisinin, an antimalarial drug, discovery fame for which she won the 2015 Nobel Prize in Physiology or Medicine), Dr Rachel Reuben (1934–2010, medical entomologist and naturalist famous largely for her contribution to the understanding of the Japanese encephalitis vector taxonomy and bioecology) and Dr Sarala K. Subba Rao (of *Anopheles culicifacies* sibling species complex fame). In the life of Dr Ronald Ross, a Major Surgeon in the Indian Military Service who won the first Nobel Prize in Physiology or Medicine (1902), very little or abysmally minute is known or written about his good wife, Rosa Bloxam and her energetic actions and devotion in silence as exhibited in all her undertakings - a marked feature that furnishes the true measure of her life.

Like all great men and women of science, art and literature, she, in true consonance of artistic habits of Dr Ross cultivated meticulously the spirit of self-help, perseverance and patience converting apparently the most insuperable proved obstacles in her way to her advantage. In fact, in many instances she turned these very difficulties her best helpers, by evoking her powers of labour and endurance, and stimulating into life faculties which might otherwise have lain dormant: "*He who can listen to the music in the midst of noise can achieve great things.*"

Rosa Bloxam had a quiet nature (Fig. 19.1). She strongly supported Ross in his undertakings during highly tumultuous I.M.S. career. Three features can be easily singled out to prove that she was a great source of inspiration and strength to Major Ross. First, during very frequent and often subtle transfer orders to Ross, she never raised a voice or tantrums even when her prospects to accompany her husband were diminished. It was an amazing sense of courage and strength Rosa Bloxam exhibited in her behavior throughout that even when soon after she landed in India, to Dr Ross's place of posting in Madras, she knew of her husband's immediate movement to Myanmar and that Ross had had to go all alone. Though Dr Ross made a comfortable stay arrangement for her in a hotel, one could understand the kind of thoughts must be crossing the mind of a young wife, new to a country several thousands kilometer away without any acquaintance or friend. But, she boldly accepted to live by herself

Fig. 19.1. Dr Ronald Ross and his wife Rosa Bloxam outside the laboratory in Calcutta. (*Source:* Photo commensurately cropped from PPT slides courtesy Col. A. Kumar).

and sent off Dr Ross on his duty. From Myanmar Dr Ross made arrangement for her to come over but hardly she had stayed there with him for some days that again he had had to move to the Andamans in the Indian Ocean for many weeks. Rosa Boxam was once again all alone in Myan Mar, another country, without husband's presence and any relative. Such a scenario persisted throughout Ross's service. Just before Dr Ross had discovered the 'mosquito-malaria connection' for the first time in the world, in India's Secunderabad, Rosa Boxam was far away in Ootacamund, staying in a hotel with their children. Ross took two months leave to stay with the family, but mostly spent time exploring the nearby Sigur and Kalputty areas for the kind of mosquito wherein the malaria parasite developed without any difficulty. His wife did not forbid him from his maddening explorations for weeks together. And, yes, Ross witnessed the first *Anopheles*, the 'dapple wings', here in Sigur which reinforced his thesis that this mosquito was the key player for developing parasite outside man's body. Apparently, they were made for each other. Secondly, Dr Ross's unhappiness with the British Indian government on account of low salary or time-bound promotion or lack of even commendations for his pathbreaking research achievements was not hidden from Rosa Boxam. But, utilizing her powers of endurance and persuasion she always guided her husband to take a well-balance step and not to jeopardize his future prospects of earning name and fame as well professional escalations. Thus, Ross never ever had either a heated conversation with any of his superior officer nor criticized any of their decisions in any matter whatso ever. And, thirdly, Rosa Boxam understood Dr Ross very well, particularly his inclination toward writing poems, novels, dramas, scientific manuscripts, official reports and personal field diaries. She was the silent source of strength for Dr Ross's literary compositions. Without her patient and quiet considerations, Ross could never have successfully accomplished a plethora of writings, leave aside his malaria research.

It is not surprising that so much little is known to the world about this wonderful person, Rosa Boxam, wife of the world famous malariologist Dr Sir Ronald Ross because he had always kept family matters away from his profession, particularly his 'do or die' subject of finding out Ariadne's thread to the centuries old malady of malaria to the advantage of millions of lives every year!

• • •

PART VII: THE INDIAN HERO

20

Dr Ronald Ross, Mosquito and Indian Psyche

होंठों पे सच्चाई रहती है, जहँ दिल में सफ़ाई रहती है
हम उस देश के वासी हैं, जिस देश में गंगा बहती है.

मेहमाँ जो हमारा होता है, वो जान से प्यारा होता है
ज़्यादा की नहीं लालच हमको, थोड़े में गुज़ारा होता है
बच्चों के लिए जो धरती माँ, सदियों से सभी कुछ सहती है
हम उस देश के वासी हैं, जिस देश में गंगा बहती है.

कुछ लोग जो ज़्यादा जानते हैं, इन्सान को कम पहचानते हैं
ये पूरब है, पूरबवाले हर जान की क़ीमत जानते हैं
मिलजुल के रहो और प्यार करो, इक चीज़ यही जो रहती है
हम उस देश के वासी हैं, जिस देश में गंगा बहती है.

जो जिससे मिला सीखा हमने, ग़ैरों को भी अपनाया हमने
मतलब के लिए अँधे होकर, रोटी को नहीं पूजा हमने
अब हम तो क्या, सारी दुनिया सारी दुनिया से कहती है
हम उस देश के वासी हैं, जिस देश में गंगा बहती है.

होंठों पे सच्चाई रहती है, जहँ दिल में सफ़ाई रहती है
हम उस देश के वासी हैं, जिस देश में गंगा बहती है.

– *(Song scripted by Shailendra and sung by Mukesh, in Raj Kapoor's 'Jis Desh Mein Ganga Behti Hai'* (1960)

The Nobel Prize to Dr Ronald Ross in 1902 "for his work on malaria, by which he has shown how it enters the organism and thereby has laid the foundation for successful research on this disease and methods of combating it" was India's first to have been won in the field of Physiology or Medicine till date. He was born to an Indian Military General Sir Campbell Claye Grant Ross who was born and brought up in India. He served the British India Army for more than three decades. He had ten children among whom Ronald Ross was the eldest. Ronald was born in Almora in the Himalaya, northern India on 13th May, 1857. He remained in India till 1865 when he was sent to England for schooling. After graduating in medicine he joined as a medical officer in the Indian Military Service in 1881, to stay and work most fruitfully in India until 1899. He returned to England in February 1899, to join there, after a brief hiatus, at the London School of Hygiene & Tropical Medicine first as a Demonstrator and then a Professor. He was there knighted and many laurels were bestowed on him. Later he became Director of the Ross Institute which position he held till his last breath on 16th September, 1932. The British government and people conferred on him several laurels, during his life or posthumously, such as for example the London School of Hygiene & Tropical Medicine by exhibs Ross' name, along with 22 other pioneers of public health and tropical medicine, on the School's Frieze, catalogues his papers in the archives of the London School of Hygiene and Tropical Medicine and the Royal College of Physicians and Surgeons of Glasgow, and much more. In 2010 the University of Liverpool has named its new biological science building "The Ronald Ross Building" in his honour. His grandson David Ross inaugurated it. The building is home to the university's facility for the Institute of Infection and Global Health. The University of Surrey, UK, has named a road after him in its Manor Park Residences. Ronald Ross primary school near Wimbledon Common is named after him, and the school's coat of arms includes a mosquito in one quarter. He was elected a Fellow of the Royal Society (FRS) in 1901 and of the Royal College of Surgeons in the same year. He was appointed Vice-President of the Royal Society from 1911 to 1913. In 1902 he was appointed a Companion of the Most Honourable Order of the Bath by King Edward VII. In 1911 he was promoted to the rank of Knight Commander of the same Order. He was also decorated with the title Officer of the Order of Leopold II of Belgium. He joined honorary membership of learned societies of most countries in Europe, and elsewhere. He got an honorary M.D. degree in Stockholm in 1910 at the centenary celebration of the Caroline Institute and his 1923 autobiography Memoirs was awarded that year's James Tait Black Memorial Prize (Gibson, 1997; Nye & Gibson, 1997). While his vivacity and single-minded search for truth caused friction with some people, he enjoyed a vast circle of friends in Europe, Asia and the United States who respected him for his personality as well as for his genius.

Even though Dr Ross is admirably known for his sanitation work in Bangalore and parasitological work on Kala azar in Assam, he will, however, is best remembered for discovering *de novo* the inextricable link between mosquito and malaria parasite (1897) and later the full life cycle of malaria parasite in birds (1898). It was the later two events which cemented his claim not only for primacy and priority of the research but also the Nobel Prize. The entire work was done in India.

Fig. 20.1. Government of India released a postal stamp in the honour of Dr Sir Ronald Ross on the occasion of centenary Celebrations on 20th August, 1997 for of his discovery of mosquito-malaria connection.

But the kind of respects, honour, recognition and identity the Indian scientific fraternity render to Dr Ronald Ross is simply unparalleled in the annals of medical history. In India, Ross is remembered with great respects as a result of his work on malaria, the deadly epidemic which used to claim millions of lives every year. There are roads named after him in many Indian towns and cities. The Government of India released a postal stamp in his honour on the occasion of Centenary Celebrations on 20th August, 1997 for of his discovery of mosquito-malaria connection (Fig. 20.1). In Calcutta the road linking Presidency General Hospital with Kidderpore Road has been renamed after him as Sir Ronald Ross Sarani. Earlier this road was known as Hospital Road. In his memory, the regional infectious disease hospital at Hyderabad was re-named as Sir Ronald Ross Institute of Tropical and Communicable Diseases. The building, located in Secunderabad near the Begumpet Airport, where he worked and actually discovered the malarial parasite is declared a heritage site named as Sir Ronald Ross Institute of Parasitology c/o Osmania University, Telangana (Fig. 20.2). The road leading up to the building SRRIP is named Sir Ronald Ross Road. In Ludhiana, Christian Medical College has named its hostel as "Ross Hostel". The young medics often refer to themselves as "Rossians".

Fig. 20.2: (a) The famous "SRRIP" at Secunderabad, and (b) Dr Ross being felicitated by a group of scientists.

In 1997 India organized an international conference in the Secunderabad-Hyderabad twin city also to celebrate 100 years of the great medical discovery of mosquito-malaria connection by Dr Ross, and, the Government of India released, among several novelties, also a postal stamp in his honour to celebrate the occasion on 20th August, 1997.

Fig. 20.3. Remembering malaria research novelties by Dr Ronald Ross at the SRRIP, Begumpet on 13th May, 2019. (*Source*: Dr B. Reddya Naik).

On 13th May, 2019 the Sir Ronald Ross Institute of Parasitology also organized a one-day long seminar (Keynote Speaker: Dr B.K. Tyagi) particularly highlighting his research explorations for the benefit of Nursing students (Fig. 20.3). On the initiation of Tata Sara Lee which manufacture a variety of anti-mosquito products, 20th August is celebrated every year as the Mosquito Day and celebrations of different kinds are organized all over the country. The Zoology Department, Osmania University, Telangana organized the 9th National Conference of Society of Medical Arthropodology titled "Emerging and re-emerging vector borne diseases: surveillance and control" at the SRRIP, Begampet on August 20-21, 2016 (*Convener:* Dr B. Reddya Naik) (Fig. 20.4).

ORGANIZING COMMITTEE

CHIEF PATRON
Smt. Ranjeev R. Acharya, IAS
I/C Vice Chancellor, Osmania University

PATRONS
Prof. E. Suresh Kumar
Registrar, Osmania University

Prof. V. Uma
Dean, Faculty of Science, OU

Prof. A. Narsing Rao,
Principal, Univ. College of Science, OU

National Conference Co-ordinator
Prof. Dr. B.K. Tyagi

Organizing Secretary
Dr. B. Reddya Naik

Joint Organizing Secretaries

Dr. C. Srinivasulu	Dr. A. V. Rajashekhar
Dr. I. Surya Kala	Dr. M. Madhavi

Dr. Nageshwar Rao Apka

ADVISORY COMMITTEE

Prof. Sugita Mathur, Head, Dept. of Zoology, OU
Prof. Geeta Rajalingam, Chairperson BoS, Zoology, OU

Prof. K. Pratap Reddy	Prof. P. Nagaraja Rao
Prof. V. Vanitha Das	Prof. S.Jithender Kumar Naik
Dr. G. Sunitha Devi	Dr. Rafath Yasmeen
Dr. Nageshwar Rao Amanchi	Dr. S. Padmaja
Dr. B. Jyothi	Dr. B. Neeraja
Dr. K. Y. Chitra	Dr. J. Venkateshwara Rao
Dr. Y. Sunila Kumari	Smt. Meenakshi
Dr. M. Radha Krishna	Dr. Juliet Vanitharani
Dr. Vijay Veer	Dr. A.C. Dhariwal
Prof. Devinder Singh	Prof. Jagbir Singh
Dr. Sajal Bhattacharya	Prof. Anita Desai
Prof. N. Chandrasekaran	Dr. Kailash Chandra
Dr. P. Jambulingam	Dr. Prafulla Dutta
Dr. Ashwani Kumar	Dr. Grace Marin
Dr. E. Pushpalatha	Prof. K. Murugan
Prof. Emmanuel Rajan	Prof. Vijaya Laxmi Sharma
Dr. K. Raghavendra	Dr. U.S.N. Murty
Dr. M. Govindraju	Sri. T. Rathna Joseph

DOCUMENTATION COMMITTEE

Mr. D. Sambashiva	Mr. M. Siddaiah
Mr. G. Devender	Mrs. B. Manjula

ACCOMMODATION

Accommodation will be arranged in the hotels and guest houses/hostels on payment basis. Please enquire in advance, preferably before, **15th June, 2016.**

HOW TO REACH SRRIP

Hyderabad is well connected by road and train from major cities in India. It is well connected by several airlines within India and from abroad.

LOCATION MAP

IMPORTANT DEADLINES

Conference Dates	: **20-21 August, 2016**
Registration	: On or before 15th June, 2016
Abstract submission	: Last date: 15th June 2016

ADDRESS FOR COMMUNICATION

Dr. B. Reddya Naik, M.Sc., Ph.D.
Professor of Zoology & Director,
Sir Ronald Ross Institute of Parasitology,
University College of Science, Osmania University,
Hyderabad, Telangana, INDIA - 500 007
Phone: 09290491044; Email: sripou@gmail.com

9th National Conference of Medical Arthropodology
on
Emerging and Re-emerging Vector-Borne Diseases: Surveillance and Control
August 20-21, 2016

Organized by
Sir Ronald Ross Institute of Parasitology & Department of Zoology
Osmania University, Hyderabad, Telangana State - 500 007, INDIA.

Fig. 20.4. Facsimile of the announcement of the 9th National Conference of Medical Arthropodology (*Source*: Dr. B.K. Tyagi)

A plaque is affixed reading "In the small laboratory 70 yards to the south east of this gate Surgeon Major Ronald Ross I.M.S., 1898 discovered the manner in which malaria is conveyed by mosquitoes" at the Presidency General Hospital, Kolkata (erstwhile Calcutta) (Fig. 20.5).

Fig. 20.5 (a-b): Ronald Ross Monument, also known as Gate of Commemoration, at the Presidency General Hospital (a) as in late 19th century (*Source*: Unknown/ content provider is CDC/Public domain), (b) as seen now (*Source*: Ragan Dutta WIKI/CC by SA 4.0),

Fig. 20.5 (c-h): (c) the Arched Main Gate, with two plaques - left, about discovery, and right, about famous poem; (d) A block after Dr Ross, (e) the Laboratory Plaque, (f) Sir Ronald Ross Laboratory, decorated with two special plaques, (g) Bust of Dr Ross (13th May, 1857 – 16th September, 1932), and (h) on the 78th Anniversary of his epoch-making discovery (*Source*: c-e, Dr S. Bhattacharya).

Very recently Ross's great granddaughter Jenny Philips, and her husband Simon Philips, visited the Ronald Ross Laboratory in the Presidency General Hospital, Kolkata, and the media covered extensively their memories about the great Dr Ronald Ross (Fig. 20.6).

যেখানে মশা, সেখানেই কদর রোনাল্ড রসের, মত প্রপৌত্রীর

অনির্বাণ ঘোষ

Fig. 20.6. Dr Ross's great grand daughter Jenny Philips, and her husband Simon Philips, visited the Ronald Ross Laboratory in the Presidency General Hospital, Kolkata, covered by the Bengali daily *Ei Samay* dt. 30th March, 2013; Courtesy Dr Sajal Bhattacharya).

The Asiatic Society honoured him with Barclay Memorial Medal in 1902 and it was handed over to him in 1903. Resolution of the Council of Asiatic Society was as follows: "*The Council of the Society awarded Barclay Memorial Medal to Major Ronald Ross I.M.S. (Retd.) in recognition of his work in the investigation of the transmission of malaria by mosquito.*"

Fig. 20.7 (a-c). (a) The world famous Asiatic Society of Bengal building, Kolkata, (b) Library and museum of the Asiatic Society of Bengal, and (c) Foundation Stone of the building. (*Source:* Dr S. Bhattacharya).

● ● ●

PART VIII: 'THE ROSS LEGACY'

21

Lessons Learnt from the Life of Dr Ronald Ross, 'The Scientist': Do not Stop Questioning – *The Real knowledge hides in Nature*

काक चेष्टा, बको ध्यानं, स्वान निद्रा तथैव च ।
अल्पहारी, गृहत्यागी, विद्यार्थी पंच लक्षणं ।।

India's Prime Minister Narendra Modi, in his address at the inauguration ceremony of 102nd Indian Science Congress in Mumbai (2015), called for "efforts to ensure that science, technology and innovation reach the poorest, the remotest and the most vulnerable person". Years ago, Jawaharlal Nehru (1937), the first Prime Minister of independent India, had also resonated the same message and professed, "It is science alone that can solve the problems of hunger and poverty, of insanitation and illiteracy, of superstition and dreadful custom and tradition, of vast resources running to waste, of a rich country inhabited by starving people... ." A national progress is the sum of individual industry, energy and uprightness, as of national decay is of individual idleness, selfishness, and vice. Her strength lies not much, as popularly believed, in its natural resources, but in the prowess of disciplined and dedicated lives of its men and women. Poverty, lack of facilities, professional displacement and descendancy, and other similar misfortunes, which invariably occur to a scientist in course of his or her scientific pursuits, are all unable to deter him from achieving his mottoes of innovations and discoveries aiming at improving the lives of his countrymen, or, for that matter, the whole of mankind. A scientist is perpetually industrious and no amount of hard work can tire him. In this way, scientists, by virtue of their inherent traits of perseverance, dedication and focused attention, take the front seat in driving the nation on its path to social security and intellectual prosperity.

Dr Abdul Kalam, former President of India and a great rocket scientist emphatically proclaimed, "India is committed to become a developed nation by the year 2020 scientists will play a dominant role in her phenomenal rise." Indian scientists, having worked by themselves, have created missions and visions for the country, igniting all along the young minds – the future of India! It must nevertheless be borne in the budding young minds of future India that a

'Developed India' is a reality in vision if only they persevere and apply themselves in accomplishing their responsibilities with full determination and dedication, like a scientist does: *"Dream is not the thing you see in sleep but is that thing that doesn't let you sleep."* In this regard the saintly advise of Maharishi Patanchali in Yogasuthra is to be remembered forever:

> "When you are inspired by some great purpose, some extraordinary project, all your thoughts break their bounds. Your mind transcends limitations, your consciousness expands in every direction, and you find yourself in a new, great, and wonderful world. Dormant forces, faculties, and talents come alive, and you discover yourself to be a greater person by far than you ever dreamt yourself to be."

(1) Lesson No. 1.: *Develop the character of "Self-Help".*

The principle of self-help, which embodies in a small compass the results of vast human experience, is the key to success of all great men and women engaged in any walk of life, but scientists in particular. The spirit of self-help is the root of all genuine growth in the individual and, exhibited in the lives of many, it constitutes the true source of national vigour and strength. Help from without is often enfeebling in its effect, but help from within invariably invigorates and energizes. This element is also inherently ingrained in the lives and works of all true scientists, who are invariably self-dependent, seek no help from others and attend to their work in all spheres of activity, however seemingly trivial, by themselves. The spirit of self-help is firmly rooted on their mettle, and absolutely nothing at all can shake their faith in this greatest of all human virtues, reminding us the most inspiring lines penned by Rabindranath Tagore in his Nobel Prize winning book Geetanjali's 'Mind Without Fear':

> *"Where the mind is without fear and the head is held high;*
> *Where knowledge is free;*
> *Where the world has not been broken up into fragments by narrow domestic walls;*
> *Where words come out from the depth of truth;*
> *Where tireless striving stretches its arms towards perfection;*
> *Where the clear stream of reason has not lost its way into the dreary desert sand of dead habit;*
> *Where the mind is led forward by thee into ever-widening thought and action ....*
> *Into that heaven of freedom, my Father, let my country awake."*

Ross exhibited invariably this trait of self-help in all his undertakings. When he decided to learn deeply microbiology he understood that learning that discipline without the aid of a microscope will not be possible. The science of microbiology was still in its infancy at that time following great works by scientists like Pasteur and Koch during the late mid-19th century, and availability of microscopes was not so common. Ross, an adventurous mind by nature, set out to manufacture one by himself, and he did it remarkably satisfactorily during the first furlough of home leave during 1888-89. This was his only microscopic tool when he discovered the oocysts on the stomach wall of a 'grey' dappled-winged mosquito in Secunderabad in 1897. Never ever had Ross waited for opportunities to emerge; he created them.

In fact, though, all men get opportunities in their scientific professional life yet not all can see them or capitalize on them. These opportunities hail so abruptly that only men with subtle alertness and scrutiny can catch hold of them to be tried out for extracting benefits:

> *"Fools walketh through forest and see no firewood,*
> *Wise have their eyes in the head."*

Ross was one who made the best possible use of any such an opportunity. All his furlough on home leave were actually spent in gaining knowledge on malaria by either creating new tools, processes, acquaintances, discussions and meetings etc. etc. He collected mosquitoes, reared them in laboratory, dissected hundreds of them, kept the plesiotypes for future identification and getting them identified by experts (mostly abroad) at a later date, preparing his own stain and process for colouring the malaria parasites for microscopic investigation, so and so forth. In essence, he did NOT lie in wait for his answers to come; he kept working hard with focused attention.

> *"Problems cannot be solved*
> *at the same level of awareness*
> *that created them."*
> – (Albert Einstein)

He worked hard to get his results which a hard-taskmaster like Patrick Manson admired very much throughout. Actually, accident does very little towards the production of any great result in life. Though sometime what is called "a happy hit" may be made by a bold venture, the common highway of steady industry and application is the only safe road to travel. It is said of Ross, that when he had already finished a manuscript in a tame, correct manner, he would get up from his desk, go out for a walk, all the while filled in his head with thoughts about his manuscript, and come back to give a brilliant finish to his work that ranks amongst the best in the world, being published in international journals like *Nature, The British Medical Journal* and *The Lancet*! But, it will not do for everyone who would produce an effect, to read in between the lines in the hope of writing an immaculate manuscript. The capability in putting these last vital touches is acquired only by the labour of a life, and the probability is, that the scientist who has not carefully trained himself beforehand, in attempting to produce a brilliant effect at a dash, will only produce a junk of a manuscript!

(2) Lesson No. 2: *A scientist must work hard, persevere and have focus.*

There is no alternative to hard work; a man can become genius by way of working hard, but if an inborn genius does not persevere or apply himself he will be doomed before long. Perseverance with focused attention is the basis for all lofty achievements in life including science. There is no set definition of the genius of a scientist, but all those who are bracketed within the specific class do have few characteristics common in them such as the continuous thin-king, focused attention, power of obsession, perseverance and application, and, on the top of all, as enunciated by Ronald Ross, 'pursue objectives and never think of success.' Fortune has often been blamed for her blindness, though the fact is it is man who is blindingly ignorant of the inexorable laws of life. We have ourselves learnt from our lives that fortune is usually on the side of the industrious, as the winds and waves are on the side of the best navigators. Thus, in the pursuit of even the most arcane and quaint scientific inquiries the commoner qualities are found the most useful, such as common sense of disciplined execution of day-to-day activities, focused attention, hard work and fullfillment of commitments etc.

A genius could boast of many things but even they cannot disdain the use of these inevitably indispensable, and yet ordinarily looking, qualities. The very greatest men, in fact, have been among the least believers in the power of genius, and as worldly wise and persevering as successful men of the commoner sort. Therefore, while some have even defined genius to be only common sense intensified, others attribute common sense being the most uncommon and rare trait which can, however, be acquired by carefully cultivating habits of reading and travelling – habits which were inbuilt in Ross's long career in the Indian Military Service. A distinguished teacher spoke of genius as the power of making efforts. Dr. A.P.J. Abdul Kalam hinted of it as being the power of lighting one's own fire – the ignited minds. Some even say that patience is genius. Indeed, there seems to be a very slight line of separation that divides the genius scientist from the one of ordinary mould. After all, great peoples do not necessarily do extraordinary things, but they all certainly do things extraordinarily!

Ross had toiled relentless for nearly two decades to unravel the mystery behind malaria transmission. He mostly lacked the basic knowledge of his major object, mosquito, or even malaria, at least in his initial days of research, did not generally have either official permission or instruments and means to carry out malaria research. On the contrary he was frequently displaced, mostly to places of no malariological interests, and at times exasperated and retributed by his superiors for his maddeningly focused attention to malaria research. Frustrated he used his leave and furloughs for malaria research, generally on the costs of his family's comforts, and investing from personal resources. A white, an English, he was not naturally accustomed to Indian environments but he dutifully braved his way through all climatic thick and thin, in sweltering heat of Rajaputana, in dusty weather of Quetta, in mosquito-ridden nights of Madras, Calcutta and Secunderabad, and the in the mountainous and sylvatic environments of Burma (Mian Mar), Andaman ils. and Assamese woods – all along keeping his objective in the forefront.

(3) Lesson No. 3: *Love research, and do it just as maddeningly as Ross did.*

It is the passion for discovering new things which keep invigorating the pursuers without any thought for any rewards; their only prized possession, if they achieve at all in one lifetime, is their discovery or invention. The road of research is often long and meandering, full of thorns, and only the determined and engrossed ones complete the journey. Even the best of opportunities and helps cannot make the idle industrious, the thriftless provident, or the drunken sober. Such reforms can only be effective by means of individual action, economy, and self-denial, by better habits, rather than by great rights. In the order of nature, the collective character of a nation will as surely find its befitting results in its life and health, as water finds its own level. In today's world some countries are often called superpowers. These countries, despite their recent histories, are both healthy and wealthy, but they have not gained this status overnight. National integrity, dutifulness, discipline and resolution were some of the finest characteristics of a large number of all those great scientists whose undaunted determination to accomplish the rarest of the tasks that helped shape the beauty of their nation.

The difference between ordinary men and scientists consists, in a great measure, in the intelligence of their observation. It is the mind that sees as well as the eye. Where unthinking gazers observe nothing, men of intelligent vision, the scientists, penetrate into the very fiber of the phenomena presented to the, attentively noting differences, making comparisons, and recognizing their underlying idea. Thus, many a great scientist before Dr. Ronald Ross had come close to make a link between the mosquito and the malaria transmission to humans, and in fact some of them were better equipped both parasitologically and entomologically including also the identification of various different species of mosquitoes, yet it was the genius of Dr Ross, a hyper-energized, combating and thinking scientist, who solved the riddle puzzling the world for many centuries because he sincerely gave value to the fact of the advice given by his mentor, Dr Patrick Manson, "Look for the flagella."

(4) Lesson No. 4: *Aim for the highest, and make yourself capable to accomplish the task.*

Aiming high in life demands investment of huge energy. Energy defines a man's inexhaustibility. It is the very central power of character in a man – in a word, it is the man himself the scientist way. Energy enables a scientist to force his way through irksome drudgery and dry details, and carries him onward and upward in every station of life. It gives impulse to his every action and soul to every effort. True hope is based on it, - and it is hope that gives the real perfume to life. It accomplices more than genius, with not one-half the disappointment and peril. There is indeed no blessing equal to the possession of a stout heart. So, when a scientist fails in his efforts, it gives a satisfaction to him to enjoy the consciousness of having done his best rather than lament the negative result. Failures, in the first place, are a teacher to a scientist who thus develops a combative skill against all odds in his humble yet determined pursuit of research, moving forward inch-by-inch anchoring to patience, and finally triumphing in his integrity. A true scientist is he who keeps walking undeterred on his courage 'even when his feet are bleeding and his limbs failing him'. What better example of combating spirit can be put forward than that of the life of Dr Ronald Ross, the world's one of the greatest and most celebrated malariologists ever.

A son of an army officer, Ronald Ross, born in Almora, India, on May 13, 1857 had no desire to join army that too in India. He nonetheless obliged his father's wish as an obedient son and entered the British Indian Medical Service, with first posting at Madras. His first impression of the country was 'full of dust, heat and mosquitoes'. Like most of his peers in the IMS he whiled away his leisure time golfing, fishing and hunting, and needing to perform his medical duties only for a couple of hours a day. The over-comfortable lifestyle of the whites in India created ripples in his mind, and, thinking of his father's grand achievements in service, he too thought to apply life to fruition. He began to think to do something extra-ordinary, something which others could never do and what could carry him on higher pedestal than others. With a pinch of salt in his day-to-day medical duties at the hands of his superiors on one or another reason, he was exacerbated to the point of being a paranoid without however indulging in any unsavory matters in either serve or social life.

Amidst the pressure of his every-day medical duties, Ross spent three years studying mosquitoes and their parasites. In doing so, he naturally faced enormous problems since he was unable to identify mosquitoes and had never seen a malaria parasite earlier. He realized that it was inevitably indispensable for him to have firsthand knowledge about the malaria parasites if he had to succeed in his objective. During his first furlough (1888), he obtained on his own costs training in public health, bacteriology and microscopy. He searched for the parasite which Dr. Laveran, a French army surgeon, had found in 1880 in the blood of victims of malaria in Algeria, but he was unsuccessful. He returned to India, with a bride and a microscope that he only had fabricated. He continued to brace failures in his attempts to see a malaria parasite or to understand any role that mosquitoes could play in malaria transmission, all this happening under the constantly increasing pressure of humiliation and indifference rendered to him by his seniors. Ronald Ross had begun to acquire a tarnished reputation of a useless medical officer who was more interested in research on malaria rather than executing his mandatory military medical assignments for which he was paid. Ross was however least interested in anything else but malaria, and soon he met with his first success of receiving an advice of "Grand induction" by Patrick Manson during his second furlough (1894) to England where he for the first time met Patrick Manson in London, who had by then already demonstrated mosquitoes' role in transmitting lymphatic filariasis and, rightly so, earned the reputation of the "Father of Tropical Medicine." It is a truism that any deeply intellectual activity in one field of science will sooner or later make its impact on other disciplines. Ross spent many hours tagging along behind Manson on ward rounds at the Seamen's Hospital and in Manson's private laboratory. In course he had impressed Manson by his inquisitiveness, ceaseless energies and the fiery desire for unravelling the centuries old mystery of malaria transmission. Manson, who then took Ross as his student, showed to Ross malaria parasites for the first time, and demonstrated to him how to find the parasites in blood smears. Manson had inferred, based on his earlier experiences in China in respect of filarial worm, *Wuchereria bancrofti* being transmitted by *Culex pipiens,* that mosquitoes could be responsible for transmitting malaria, too, although his hypothesis remained merely a conjecture as he had himself proved unsuccessful in bringing his point home.

One November afternoon in 1894 Manson began to expound upon his ideas about the plasmodia. "Do you know," he remarked to Ross while the two were on the way to the hospital, "I have formed a theory that mosquitoes carry malaria just as they carry filaria." Though this nicely rhymed observation might have been made into another new poem, Ross was initially unimpressed. The mosquito had been sugge-sted as a carrier of malaria many times before, he responded. But then the two men discussed the facts that favored this argument. As Ross began to be swayed, Manson brought in his own theories. The filaments in the crescents, believed by everyone to be evidence of a death spasm, were actually living bodies. The mosquito, feeding on the blood of a malaria patient, sucked the filamented crescents into its stomach. The filaments proceeded to travel through the stomach into the insect's tissues. After the mosquito died laying its eggs, the "flagellated spores" emerged into the water, ready to infect anyone who came to drink. These theories, which had earned for Manson the titles "pathological Jules Verne" and "Mosquito Manson," sent the young Ross into raptures of ecstasy. Suddenly the fame that had eluded him despite years writing poems, music, plays, novels, and equations seemed within his grasp. He had but to prove what Manson had presented to sound like gospel truth. Overwhelmed by the conviction that Manson was entirely correct, Ross made only the most superficial attempt to familiarize himself with the life cycle of the mosquito, despite Manson's entreaties. After having failed to examine books on the mosquito in the library of the British Museum, the only place in England at the time with such books, Ross later confessed that he was "very badly equipped for the fray." With a dream in his eyes, Dr. Ross, equipped with the resolution to prove Manson's "Grand Induction" of mosquito-malaria parasite human linkage, returned to India once more. His ignorance was offset by an almost manic enthusiasm. On the ship to India, Ross rushed among passengers and crew members, frantically pricking fingers and examining blood. At the ship's ports of call, he besieged the local hospitals for blood specimens of malarious patients. He even tried to prepare himself for anatomical studies of Indian mosquitoes by dissecting the ship's cockroaches!

Back in India, at his new posting in Secunderabad, Ross continued to dart around like a frenzied mongoose looking for cases of malaria. His preoccupation quickly incurred the hostility of his boss in the IMS. Even the natives, their fingertips inflamed from his repeated pinpricks, would run away to hide from him. Colleagues thought him so mad that they kept authenticated malaria cases from him. But he refused to give up. Spurned in his garrison hospital, he haunted the municipal hospitals and the other regiment infirmaries looking for cases of malaria. The local doctors would often turn over these cases to him just to rid themselves of his annoying presence. He offered the malarious natives a rupee a prick, yet many still fled. After he finally managed to obtain the cooperation of one patient with a documented case, fellow doctors challenged him to demonstrate the parasite. But somehow, when the moment came, the plasmodia mysteriously vanished from the blood. Poor Ross had to endure gales of laughter.

Ross bean to get frustrated and tensed by the repeated failures. He had only one hope to hang on, Manson. Following Manson's instructions, Ross captured the ubiquitous mosquitoes and tried to induce them to bite malaria patients. But they obdurately refused to bite anyone, even Ross himself. The mosquitoes caught were probably too frightened to bite, Ross reasoned. So he raised new mosquitoes from grubs. Still no luck. He baked the patients in the hot sun "to bring their flavour out." Nothing happened. On May 13, 1895, Ross's birthday, a heavy rain soaked the bed and netting of a malaria patient. Made ravenous by the moisture, the mosquitoes attacked the patient with alacrity. Ross grabbed four of them, expressed their ingested blood on a glass slide, and peered at it with his microscope. Just as Manson had professed, there were the parasites. To be certain of the results, Ross tried the same experiment with six more mosquitoes the next day. Abjuring food and sleep, he peered through the microscope so intently that his eye was blackened by repeated contact with the eyepiece. The parasites are present in the blood of the mosquito, and what is even more, they appear to be there in greater numbers than in blood from the finger. Also, the development of the crescents, and the formation of the flagella, seem to be favoured by conditions in the mosquito's stomach. The crescent-sphere-flagella metamorphosed inside the mosquito to a much greater degree than in control specimens of finger blood. At one point, Ross became so enthralled watching a white blood cell try to engulf a flagellum that in his letter he compared the battle to a swordfight. "I shall write a novel on it in the style of The Three Musketeers," he added joyously. Ross apparently drew energies from his artistic skills!

Dr. Ross, typical in his military-disciplined daily chore of maintaining minute-to-minute account of malaria-mosquito explorations, vigorously corresponded with Dr. Manson in England and sought his advice. Manson immediately wrote back with more instructions. "Let mosquitoes bite people sick with malaria," he advised, "then put those mosquitoes in a bottle of water and let them lay eggs and hatch out grubs. Then give that mosquito-water to people to drink." So Ross allowed four mosquitoes to feed on a patient named Abdul Kadir. These insects were then kept in a bottle full of water until they died. After the promise of a suitable emolument, Lutchman, Ross's native servant, was persuaded to swallow the liquid in the bottle. Day after day for ten days, Ross waited anxiously for Lutchman to develop fever. On day eleven, Lutchman complained of a headache and was found to have a slight temperature elevation. Now thoroughly excited, Ross admitted his experimental subject to the hospital and impatiently sat by him, measuring his temperature every thirty minutes. Lutchman probably only had the flu. Not a parasite was to be seen in his blood. A few days later, he had recovered completely. Ross repeated the experiment with other volunteers, but these men were likewise totally unaffected by the mosquito water. The volatile Ross was plunged into deep despair. Nothing seemed to be going right. Careful dissection of the mosquitoes revealed that the flagella Manson believed to be spores remained active only a few moments in the insect's tissue. Perhaps, thought Ross, ingestion of the parasites by the mosquitoes was only incidental, unrelated to human disease. Thoroughly discouraged, he began, as usual in his moments of dismay, writing poems and novels, thinking perhaps this was a better way to succeed.

As a responsible teacher, Manson desperately tried from England to re-energize his slacking convert. "Above everything, don't give it up. Look upon it as a Holy Grail and yourself as Galahad and never give up the search, for be assured that you are on the right track. The malaria germ does not go into the mosquito for nothing, for fun, or for the confusion of the pathologist. It has no notion of a practical joke. It is there for a purpose and that purpose, depend upon it, is its own interests; germs are selfish brutes." The Indian Medical Service did not share Manson's views when Dr. Ross tried to convince them of his intentions to pursue such a research. To them Ross' search seemed more of an unholy nuisance than a Holy Grail. He was transferred to Bangalore to tackle cholera epidemic, and there he was neck-deep busy for the next eighteen months and had little time to do anything but visit cholera cases and fight with civilian and military authorities over sanitary reforms. In England, Manson too was distressed for the official indifference towards Ross's works on malaria. Perhaps someone else who would give him no credit might appropriate his precious theory. "The Frenchies and Italians will pooh-pooh it, then adopt it, and then claim it as their own," he warned Ross in one letter, "See if they don't. But push on with it, and don't let them forestall you. They won't have this autumn, and they will not have a chance to work seriously at the matter until next June or so. You have got a year ahead of them." The roadmap was clearly laid open before him; prepare or perish.

All along these duties in Bangalore his family lived away in Ootucamund. So Ross went on two months' leave whereupon Ross tried to continue his malaria work on his own expenses. But within a few days he himself contracted malaria and spent his vacation too ill to do anything. When he finally recovered, the weather had turned hot and dry. A stiff wind filled his laboratory with mounds of dust. The sweat-rusted microscope, its one remaining eyepiece cracked, sat idle as Ross sulked, brooded, and wrote, his all-time accomplice and refuge, poetry. As he was emerging from this bleak period of depression, Ross was struck by the crucial inspiration. Re-reading Manson's original article on filariasis, he was reminded of the fact that only one species of mosquito among the four found in Amoy, *Culex pipiens*, was capable of carrying filariasis. Dr. Ross was by now forty years old but, though he was well known in India, both for his sanitary work at Bangalore and for his researches on malaria, he received no advancement at all for his pains. Still enthusiastic, he communicated his new notion to Manson who, however, had quite an imperfect understanding of mosquito behavior. Believing that the insect bit only once during its life, he was convinced that Ross could not be correct. "Follow the flagella," he wrote back, "and forget this crazy idea." Ross obediently went back to dissecting mosquitoes and in February 1897 was able to observe the true fate of the flagella. Within a blood smear he saw two parasites near each other. The first was giving off flagella, while the second, which was spherical and unsegmented, had a single flagellum wriggling slowly inside. Had Ross begun a moment earlier, he might have watched the flagellum emitted by the first parasite penetrate the second and so perceive the true nature of the process. But from his vantage point, he surmised that the single wriggling flagellum was trying to escape the sphere rather than fertilize it. When Dr. McCallum in Baltimore, USA, correctly interpreted the process a few weeks later, Ross felt deeply humiliated and disgraced as a man of science for incorrectly interpreting his own observations.

Ross was by this time entirely engrossed in his malaria research and was convinced that the study of malaria was to be his path to fame and fortune. He pestered the military hierarchy incessantly to allow him time for research. But he had stepped on too many toes to expect any favours from his superiors. In June 1897 he was sent back to the steamy, depressing hospital in Secunderabad. Yet, he refused to give up. Manson had also suggested that each form of the malarial plasmodia might require a particular mosquito species. Ross suddenly realized he had used the wrong species of mosquito. Most of his cases had been falciparum malaria, and he had consistently employed the common house mosquito, a brown insect with plain wings that sat with its tail pointing down. The larvae of these mosquitoes floated head downward in water, breathing by means of a long tube. After a chance encounter he encountered an altogether different type of mosquito, the "dapple wings" as he called it, whose adult sat with its tail pointed upward and larvae had no breathing tube and floated parallel to the water surface. A malaria patient named Husein Khan was the first experimental subject. Ross let loose a dozen of the grey mosquitoes under the mosquito netting of Husein Khan's bed, and trapped each one, after it had fed, in a separate bottle. Two mosquitoes were then examined immediately. Two days later others were dissected. More of the insects died during the night and quickly decomposed. In all of them Ross found nothing. Worse, search as he might, no more of the same species of insect were to be procured. On August 20, 1897, a tired, discouraged Ronald Ross dissected one of the two remaining brown mosquitoes. Just as he was about to give up, he noted a queer structure within the cells lining the insect's stomach: an almost perfectly circular cell containing a group of black pigment granules very similar to those seen in the malarial parasites found in blood smears. Nearby were identical circular cells. Ecstatic with joy, Ross repeated dissection two more mosquitoes on August 21 and again saw the pigment-filled circles. He jumped to his feet with ecstasy and wrote to Manson, "I really believe the problem is solved." Manson took forward his great discovery to learned academics and researchers in the European and Western circles. This later came up as a proof of priority of Ross's original discovery of mosquito-malaria connection – the basis of award of the Nobel Prize!

(5) Lesson No. 4: *Take your results to fruition leaving no stone unturned.*

There are a few scientific phenomena and their discoverers with which we are so familiar that it appears impossible that their truth could ever be doubted. It seems incredible that only a few decades divide us from the discovery of the principle of the inexplicable link between mosquito and malaria. Such was the indelible imprint of the discovery of the mosquito-malaria connection that it overthrew the accepted doctrine of a century and proved what earlier broods of scientists had thought about the bad air and malaria disease or even dead-mosquito polluted water's link with malaria in humans was utterly baseless and wrong. Ross's epoch-making 'mosquito-malaria connection' discovery was the starting point of Ariadne's thread that would eventually lead to the exit from the labyrinth of human malaria and lay the foundations for scientifically based disease eradication through vector control in future. Truly his discovery, or rather his observation, ushered in a new era of medical science. The road ahead was, however, not easy and was full of difficulties, particularly precocious displacement. He was denied any assistance sought to carry out next steps of his malaria research. Yet, he was calm and resilient, and kept his focus at every cost:

> *"Real glory,*
> *Springs from the silent conquest of ourselves,*
> *And without that the conqueror is naught,*
> *But the first slave."*

Though Ross was exceedingly ecstatic about his discovery, but he also within his heart knew that there were many questions that remain to be yet answered, particularly the peregrination in the mosquito body the malaria parasite would undergo from the stomach stage of oocysts onwards to get ready to be released as infective stage in the human body! By this time the repeated failures had prompted Ross to question Manson's theory and he began to formulate his own hypothesis that the mosquito needed to inject a small quantity of fluid into the human body with each of her bite. To test this eminently reasonable idea, Ross allowed mosquitoes that had fed on a malaria patient to bite a healthy man. Nothing happened. The experiment was repeated again and again. No fever developed.

Meanwhile he persuaded Manson to intercede with high government figures on his behalf. Manson thereupon approached Lord George Hamilton, the secretary of state, and even petitioned Lord Elgin, the Viceroy of India, but of no avail! As luck would have it, swarms of grey mosquitoes suddenly appeared in Secunderabad. Now able to repeat his experiment, Ross confirmed that only those insects that had bitten malaria patients carried the circular cells. He submitted the results to the British Medical Journal, and in December 1897 his report, "On Some Peculiar Pigmented Cells Found in Two Mosquitoes Fed on Malarial Blood," was published. Then, just at his moment of triumph, he received his movement orders: UNDER INSTRUCTIONS FROM COMMAND HEADQUARTERS, SURGEON MAJOR R. ROSS, I.M.S., WILL PROCEED IMMEDIATELY TO BOMBAY FOR MILI-TARY DUTY. Infuriated, Ross bombarded the commanding officers with angry telegrams. Even Manson's intervention did not help. His new station, to his dismay, was totally unsuited to his research.

Yet, the resourceful Ross did not quite languish. He knew of Danielewsky's studies of bird malaria, and verified for himself that some types of pigeons carried the disease. At the time, many biologists believed that mosquitoes did not attack birds. After studies of pigeons, sparrows and crows, Ross verified that birds were indeed bitten by mosquitoes, as well as by other insects. In the meantime, Manson reached the right bureaucrat and on January 29, 1898, after weeks of futile pleading for a furlough or a change of station, Ross was notified by wire of his attractive new assignment in Calcutta (now Kolkata) In almost no time Ross could demonstrate that the plasmodium passed from bird to mosquito, just as it did in humans. Moreover, the same pigmented, circular cells would form in the wall of the insect's stomach. Only one difference was noted. The common brown mosquito (*Culex*) was the carrier of bird malaria. The grey or dappled-winged vector (*Anopheles*) of human malaria could not be infected by the bird parasite. These studies were carried on in desperate haste. Ross was forever haunted by the fear that hostile superiors would once again interrupt his work. He was forbidden to publish any of his results, though compelled to send long, meticulously detailed summaries to headquarters. The only outsider aware of the research was Dr. Patrick Manson - his protector, advertiser, and mentor.

But how did the parasite pass from the mosquitoes to the birds? According to Manson's theory, the parasites were ingested with water in which the mosquitoes had died while laying eggs. Ross easily tested this theory by feeding infected mosquitoes to healthy sparrows. The result: The birds remained free of malaria. The true answer finally emerged as Ross continued to study infected mosquitoes. As the circular cells within the mosquito's stomach enlarged, the pigmented granules grew into little rod-shaped bodies. And soon, Ross discovered, these rods appeared in the insect's thorax. Since the stomachs of some mosquitoes showed only ruptured capsules where circular cells had been, the implication was clear: the rods migrated. On July 4, 1898, Ross discovered their destination. Examining the insect's head, he noted the salivary gland to be so loaded with rods that it quivered. Here then was the answer. Malaria was passed back to the birds through the mosquito's saliva during the act of biting. As a final verification, he sent Mohammed Bux to capture a group of healthy sparrows. Mosquitoes that had fed on infected birds were allowed to bite these healthy ones. Within a few days the blood of the new birds was loaded with malarial parasites. Ross communicated his results to Manson in a state of intense excitement. Manson received the news at a meeting of the new Tropical Diseases Section of the British Medical Association. When he read Ross's report to the assembled delegates, the excitement generated was intense. "I am sure you will agree with me," Manson added, "that the medical world, I might even say humanity, is extremely indebted to Surgeon Major Ross for what he has already done, and I am sure you will agree with me that every encouragement and assistance should be given to so hard-working, so intelligent, and so successful an investigator to continue his work." The audience might have agreed with Manson, but the Indian Medical Service did not. For Ross's original assignment was also supposed to include the study of another tropical disease: Kala-azar. And so, just at the peak of his success, Ross received instructions to abandon the malaria work and report to a new post in Assam to do research on Kala azar. In a state of deep despondency, Ross was forced to close his small laboratory in Calcutta and leave for his next destination. In fact, the research still to be done was more complex than that. For, as Manson remarked in his address to the Tropical Diseases Section, "One can object that the facts determined for birds do not hold necessarily for men."

Ross had a tumultuous career throughout since many of his medical superiors and colleagues did not see eye-to-eye. When his service bosses were unsympathetic to his requests for leave, at times he threatened to take early retirement from the I.M.S. and return to his first love, namely, literature. It was the rock-like support of Manson, his mentor till that time, who, on one hand, encouraged him to continue his research during these periods of discouragement and despair, but, on the other hand, also endeavoured hard to convince his I.M.S. bosses to think favourably about his efforts. For his pioneer work, Ross was awarded the 1902 Nobel Prize in Physiology or Medicine. He subsequently held various positions notable amongst which were (i) Professor of Tropical Medicine at University College, Liverpool (from 1902), (ii) Physician at King's College, Hospital, London, (iii) Consultant on malaria to the War Office, and (iv) Director of the Ross Institute and Hospital for Tropical Diseases, London (from 1926). He was knighted in 1911.

"The world is his
who has patience and industry."

(6) Lesson No. 6: Have confidence in yourself.

Development of research findings has a queer tendency to evolve almost simultaneously in more than one place, as perhaps best exemplified by the independent rediscovery of tenets of genetics, originally propounded by Gregor Mendel in the mid-1800s, in three countries in the early 1900s; the discovery claims for the Human Immunodeficiency Virus (HIV) causing Acquired Immunity Deficiency Syndrome (AIDS) in at least two countries, France (whose scientists took away the Nobel Prize after a bitter battle of priority and primacy of research) and the United States of America, in the 1970s. In the similar manner, during the late 1870 and early 1880s, mosquito had begun attracting much attention of tropical disease researchers and Manson's assertion of the mosquito's role in transmitting human lymphatic filariasis worked to further ignite subject ebullience at a feverish pace all over the world. Amongst all the vector-borne diseases being pursued at that time, the attraction which malaria commissioned was unparalleled in the entire history of tropical diseases around the globe. Malaria, for some known reasons, always asserted some claim to be special. Widespread and debilitating malaria – the most common of all tropical infections – was of all-time great interest, as scientists in France, Italy, Germany, United States of America, England and India vied in their independent and isolated pursuit with a rather nationalist fever for research excellence and discovery claims. The struggle was no less tense than the sprint to discover the New World or the 'Golden bird' that India had once been, and it foreshadowed future scientific contests over the polio or small-pox vaccine and the atom. Amongst a large number of much acclaimed and seasoned names in circulation, the Italian genius Professor Giovanni Batista Grassi was already a hugely respected scientist with a unique combination of knowledge in both medicine and zoology, not to mention in the least his vast prowess in animal taxonomy including mosquitoes, ably aided with a background of several years mosquito-malaria research, much in advance to that of Dr. Ronald Ross in India.

Grassi (1854 – 1925) obtained his doctorate in 1878 and became Professor of Zoology at the University of Catania till 1996. He demonstrated (1898) that the *Anopheles* mosquito carried the *Plasmodium* parasite of malaria in its digestive tract. He was known also for his researches on the life history and societies of the Termitidae, and on the developmental relationship between *Leptocephalus* and the common eel and other Muraenidae. Unlike the mercurial, yet fumbling Ross, Grassi was as precise, unemotional, and meticulous as a watchmaker. His scientific papers were models of perfection; classics in their own time that he would labour over for years before publishing. Though his eyesight was terrible, he had done studies of the microbes that parasitized white ants or termites and of the behavior of eels. He was a celebrated mosquito taxonomist of his time and readily realized the mosquito-malaria theory of Manson. He had also cleverly guessed that only certain mosquitoes were involved in the malaria transmission. Further, since mosquitoes were numerous in some locations where malaria was rare, while in other places malaria was common and mosquitoes rare, he deduced that only a particular mosquito could carry the disease. And no one was better qualified to identify this insect than Grassi, a superb taxonomist capable of recognizing each of the more than thirty species indigenous to Italy at that time.

In the summer of 1898 Grassi began his search. Using large test tubes, he trapped every variety of mosquito that he spotted. At the same time, he recorded in a notebook the incidence of malaria at the place where each insect had been captured. After a summer spent visiting most of the swamps in Italy, Grassi was able to exonerate all forms of the *Culex*, or common brown house mosquito, the insect that caused Ross to throw away almost two years' work. The peasants called the type that appeared to be responsible zanzarone; it had dappled wings and a tail that, pointed upward when it sat, Ross's "grey mosquito." In September 1898 Grassi was able to report that this insect, *Anopheles claviger*, was the carrier of human malaria. The proof was obtained by means of a human experiment. The subject, a Mr. Abele Sola, had been for six years a patient in the Hospital of the Holy Spirit, a building high atop one of the hills of Rome. Malaria had never been seen in this vicinity and neither had the anopheles mosquito. With Mr. Sola's permission, Grassi and Dr. Giuseppe Bastianelli, a hospital physician, shut Mr. Sola in a room with *Anopheles* mosquitoes every night for ten nights. On the eleventh day, the patient developed a malarial chill. Examination of his blood revealed large numbers of plasmodia: "The rest of the history of Sola's case has no interest for us," Grassi wrote, "but it is now certain that mosquitoes can carry malaria, to a place where there are no mosquitoes in nature, to a place where no case of malaria has ever occurred, to a man who has never had malaria, Mr. Sola!"

Grassi's successful repetition of the experiment on other patients somehow leaked out. The newspapers were incensed and implied that Grassi was ruthlessly endangering the lives of his human guinea pigs. He ignored them and continued with his work. For there was still Robert Koch's assertion that the infant anopheles inherited the malarial infection from its mother. To refute this, Grassi raised anopheles mosquitoes from eggs. Then he allowed the mosquitoes to bite him and six other volunteers. All remained free of malaria. In the course of his experiments, Grassi reportedly came to read Ronald Ross's articles on bird malaria. After dissecting the *Anopheles*, Grassi was able to observe the same pigmented circular cells and the same rods within the salivary gland. But when he published, he failed to give Ross credit. Needless to say, Ross was furious. Thoroughly convinced that Grassi was trying to steal his discovery from him, Ross sent angry letters to the journals that had published Grassi's papers, asserting that Grassi was a mountebank, a cheap crook, a parasite who survived on the ideas of others. Grassi replied in equally acrimonious terms. So vicious did the correspondence become that journal editors, fearful of libel, hesitated to publish the letters. But Ross and Grassi did not stop feuding. Both enlisted the aid of the authorities on tropical medicine. Ross was able to obtain letters from Dr. T. Edmundston Charles, an English observer of the Italian work in Rome. Using this evidence, Ross asserted that Grassi had been aware of the studies on bird malaria, though Grassi later denied such awareness. When Ross could not find a publisher for a book containing his case against his Italian adversary, he paid for the printing himself, carrying the work through two editions. This bitter conflict lasted for more than two decades.

In the meantime Grassi began a campaign against malaria, since opposition to the mosquito theory was still widespread. He warned against taking walks in the twilight, the prime mosquito feeding time. "Don't go out in the warm evenings," he announced, "unless you wear heavy cotton gloves and veils." Naturally he was laughed at. To prove his point, Grassi set up an experiment to prevent malaria in the most heavily diseased region in Italy, the railroad line that ran through the plain of Capaccio. With funds from the queen of Italy and authority from the railroad, Grassi installed fine mesh screens on the doors and windows of ten station-masters' houses. One hundred twelve employees were paid to stay inside during the twilight. Another four hundred fifteen workers went out as usual. At the end of the summer, almost all the unprotected developed malaria. But of the hundred and twelve protected individuals, only five got sick. "In the so much feared station of Albanella," wrote Grassi triumphantly, "from which for years so many coffins had been carried, one could live as healthily as in the healthiest spot in Italy!"

Professor Grassi's epoch making research impressed Dr. Patrick Manson in England who repeated Grassi's experiment in an even more dramatic fashion. During one phase, Manson sent men from the staff of the London School of Tropical Medicine to live and work in the Roman Campagna. One hour before twilight they repaired to special mosquito-proof tents, where they stayed until an hour after sunrise. Though everyone around them contracted malaria, these men stayed healthy. This outcome was dramatized by the fate of a police detachment sent from Rome to capture a criminal in the Campagna while the Englishmen were there. Though they remained in the Campagna for only a day, all the policemen developed malaria shortly after returning to Rome. The second phase of Manson's experiment was most chilling, though it ended happily. Manson managed to obtain live, malarious Italian mosquitoes, which he permitted to bite his own healthy son. In fourteen days, the boy was ill with malaria. Manson's laboratory assistant, George Warren, then allowed a few more of the infected mosquitoes to bite him, remarking that it would have been "a great pity to waste them." He too quickly contracted, malaria. Both young men survived. After these results had been reported in newspapers and magazines throughout the world, the last resistance to the mosquito theory finally crumbled.

Even though the balance of pioneering research work on Anopheles mosquito as the carrier for malaria weighed heavier on Ross's side, nevertheless no one could ever deny the genius of Professor Grassi, whom many considered as the most authoritative zoological scientist of his time. Dr Patrick Manson, who mentored Ronald Ross and played the key role in his achieving the most significant discovery of the 19th century, was even in favour of Grassi when a decision about the 1902 Nobel Prize was to be taken! The Italian Government had duly recognized the genius, however, and issued a stamp (25 L) in his honour on November 19, 1955, on the occasion of his 30th death anniversary, bearing portrait of Battista Grassi, mosquito and microscope. The people of Italy also dedicated a boulevard in Milan in his memory and named it as Giovanni Battista Grassi Street 74, harbouring the famous The Luigi Sacco Hospital. Professor Grassi received many honours and distinctions among which the Darwin Medal presented to him in 1896 was very dear to him.

None of the two individuals, Ross and Grassi, was lesser in any sense of quality research and both were full of knowledge, wisdom, energy, passion and determination. Like in every generation a few individuals toil and struggle to achieve seemingly impossible feats, the two of these extraordinary rose against all odds, even when their own survival was not assured. They never gave up, or accepted defeat. Their lives inspired and beckoned people from successive generations to take up challenges, to rise above drab mediocrity, and to excel. Life of every such individual is unique even though there may be some commonalities in the circumstances from which such individuals rise. No society can progress without role models. Precedents of successes from extremely harsh conditions of both Professor Grassi and Dr Ross motivated many who had given up the struggle. Giovanni battista Grassi was such a man. Even though Ross and Grassi locked their horns for most part of their lives and did not see eye to eye with each other at all, nevertheless both had also that unsaid lofty regards for each other's scientific marvels! Regardless of the outcome of their personal skirmishes, one thing which was common in them was that they both were endowed with seemingly limitless self-confidence, energy, patience and optimism that are so important characteristics of a true scientist.

In essence, the young Indian scientist, or for that matter one anywhere in the world, but especially a medical entomologist/arthropodologist, has many lessons to be learnt from the life of Dr Ronald Ross. Yet, the most indispensable lesson is that to inevitably take laboratory knowledge out in the open field for final testing since it is the Nature which is the most powerful classroom for all worldly wisdom and embodies within its varied ecosystems answers to all kinds of questions.

• • •

References Cited

Bedley, Brenton Thoburn, 1931. *Visions and Victories in Hindustan: A Story of the Mission Stations of the Methodist Episcopal Church in Southern Asia*. Volume 2. Methodist Publishing House, 842 pp.

Bhattachaya, S. and S. Bhattacharya, 2010. Ronald Ross: Known Scientist, Unknown Man. *Science and Culture* 76(7-8) 253-256.

Biesel, Uli Beisel and Boete Christophe, 2013. The Flying Public Health Tool: Genetically Modified Mosquitoes and Malaria Control. *Science as Culture* 22 (1): 38-60.

Bynum, W.F. and C. Overy, 1998. *The Beast in the Mosquito: the Correspondence of Ronald Ross and Patrick Manson*. Amsterdam-Atlanta, GA, Editions Rodopi B.V. 5-520.

Capanna, E., 2012. Grassi versus Ross: who solved the riddle of malaria?. *International Microbiology* 9 (1): 69–74.

Chernin, E., 1988. Sir Ronald Ross, malaria, and the rewards of research. *Medical History* 32: 119-141.

Dutta, Agneish, 2009. Where Ronald Ross (1857–1932) worked: the discovery of malarial transmission and the Plasmodium life cycle. *Journal of Medical Biography* 17(2): 120–122.; DOI: 10.1258/jmb.2009.009004

Ghosh, Amitav, 2001. The Calcutta Chromosome: A Novel of Fevers, Delirium & Discovery. Penguin India, 270 pp.

Gibson, M. E., 1978. Sir Ronald Ross and his contemporaries. *J. R. Soc. Med.* 71: 612;

Gibson, M. E., 1983. A catalogue of the Ross Archives, London School of Hygiene and Tropical Medicine. [text and microfiche], p. 16.

Gibson, M.E., 1997. Ronald Ross and India. *Indian J. Malariol.* 34:56-70.

Harbach, R.E. and I.J. Kitching, 1998. Phylogeny and classification of the Culicidae (Diptera). *Systematic Entomology* 23: 327–370.

Harrison, G., 1978. Mosquitoes, Malaria and Man: A History of Hostilities Since 1880. J. Murray. London.

Knight, Kenneth L., 1972. History of Mosquito Systematics: Part I. Eighteenth Century. *Mosquito Systematics* 4(l): 10-15.

Laveran, A., 1880. Note sur un nouveau parasite trouvé dans le sang de plusieurs malades atteints de fièvre palustres. *Bull Acad Med.* 9:1235–6.

Laveran, A, 1884. Traitè des Fièvres Palustres. Doin, Paris.

Lovett, Richard, 1899. The history of the London Missionary Society, 1795-1895. Princeton, New Jersey.

Linnaeus, C., 1758. *Systema Naturae.* Regnum animale, ed, 10.

Manson, P., 1878. On the development of *Filaria sanguis hominis* and on the mosquito considered as a nurse. J. Linn. Soc. (Zool.) 14: 304-311.

Nye, Edwin R., Mary E. Gibson, 1997. *Ronald Ross: Malariologist and Polymath: a biography.* Basingstoke, McMillan.

Reinert, J.F., R.E. Harbach and I.J. Kitching, 2004. Phylogeny and classification of Aedini (Diptera: Culicidae) based on morphological characters of all life stages. *Zoological Journal of the Linnaean Society* 142: 289–368.

Reinert, J.F., R.E. Harbach and I.J. Kitching, 2006. Phylogeny and classification of Finlaya and allied taxa (Diptera: Culicidae: Aedini) based on morphological data from all life stages. *Zoological Journal of the Linnean Society* 148: 1–101.

Ross, R., 1897a. On some peculiar pigmented cells found in two mosquitoes fed on malaria blood. British Medical Journal 2: 1786-1788.

Ross, R., 1898a. Further observations on the transformation of crescents. Indian Medical Gazette (July, 1898).

Ross, R., 1898b. Preliminary report on the infection of birds with Proteosoma by the bites of mosquitoes. Government Press Calcutta (11 Oct, 1898); Indian Medical Gazette (Jan. 1899); British Medical Journal (18 Feb., 1899).

Ross, R., 1904. *Researches on malaria. The Nobel Medical Prize Lecture for 1902.* Stockholm, Sweden.

Ross, R., 1923. *Memoirs, with a full account of the Great Malaria Problem and its solution.* John Murray, London. 547 pp.

Spielman, A., Antonio D. Mosquito, 2002. The Story of Man's Deadliest Foe. Hyperion, 2002, 272.

Power, Helen J., 1998. *Tropical Medicine in the Twentieth Century: A history of the Liverpool School, 1898-1990.* London, Kegan Paul.

Pradhan, M., 2009. Life sketch of Dr. Ronald Ross. Journal of the Asiatic Society Vol LI (2): 63-66.

Tyagi, B.K., S.K. Bansal and K.V.S. Verma, 1991. Intra-specific morphological variations in two ecotypes of *Anopheles Stephensi* Liston in north-western Rajasthan (Dipter: Culicidae). *Annals of Entomology* 9: 85-89.

• • •

ANNEXES

Annex 1	:	Malariological bibliography of Dr Ronald Ross
Annex 2	:	Non-malariological bibliography of Dr Ronald Ross
Annex 3	:	New species of parasites discovered/described by Dr Ronald Ross
Annex 4	:	Ph.D./M.Sc. Theses about Dr Ronald Ross
Annex 5	:	Post-retirement from British Indian Medical Service important positions/assignments /consultancies rendered by Dr Ronald Ross
Annex 6	:	Laurels, distinctions, awards, prizes and recognitions to Dr Ronald Ross
Annex 7	:	Significant prizes and awards won by Dr Ronald Ross
Annex 8	:	Fellowship and Honorary memberships of Scientific societies conferred on Dr Ronald Ross
Annex 9	:	Literary works by Dr Ronald Ross
Annex 10	:	Dr Ross, the mathematician
Annex 11	:	Books by Dr Ronald Ross
Annex 12	:	Unpublished reports by Dr Ronald Ross
Annex 13	:	Works of Dr Ross on parasites other than malaria
Annex 14	:	Autobiography by Dr Ronald Ross
Annex 15	:	Places in India where Ross worked
Annex 16	:	Annotated catalogue of extant species in India

Annex 1. **Malariological Bibliography of Dr Ronald Ross**

Jones, R. Ross and Ellet, 1908. La malaria fattore transcurato nella storia di Grecia e di Roma. Napoli Detken e Rocholl.

Ronald R., 1918a. An interim report on the treatment of malaria: Abstract of 2460 cases. War Office Publication, with discussion, pp. 342; Trans. Soc Trop. Med. & Hyg (Mar-Apr, 1918), Vol. 10 (No. 5 and 6).

Ross, R., 1893a. Some objections to haematozoic theories of malaria. Medical Reporter, 2, pp. 65-71.

Ross, R. (1893d) 'The chest measurements of recruits', IMG, 28, pp. 4—5.

Ross, R. (1893c) 'Nodulated and vacuolated corpuscles', IMR, 4, pp. 213-4.

Ross, R., 1896e. Some experiments in the production of malaria fever by means of the mosquito. South Indian Branch, British Medical Gazette (Dec. 1896).

Ross, R., 1897e. The remarks on Twelve Natural Developments etc. International Congress of Hygiene, Budapest, 1894. British Medical Journal.

Ross, R., 1898e. Report on the cultivation of *Proteosoma*, Labbé, in grey mosquitoes. *Indian Medical Gazette* 33:401-448.

Ross, R., 1899e. Inaugural Lecture on the Possibility of Extirpating Malaria from Certain Localities by a New Method. *British Medical Journal* 2: 1-4.

Ross, R., 1899j. Report on Kala-azar. Govt. Press, Calcutta (30 Jan 1899).

Ross, R., 1902e. Evidence regarding the discovery of serum diagnosis. Lancet 1

Ross, R., 1902f. Deutsch. Med. Woch. (27th March, 11902), S. 231.

Ross, R., 1903e. Papers on the thick-film process for malaria diagnosis. *Lancet* (10th Jan., 1903).

Ross, R., 1905e. The logical basis of the sanitary policy of mosquito reduction. *Science* 22: 689-699.

Ross, R., 1906b. Malaria in Greece. J Trop. Med. (15 Nov., 1906).

Ross, R., 1906e. Medical Society, Oxford.

Ross, R., 1907e. Malaria campaign. The prevention of malaria in British possessons. [Berichtuber den xiv en Internationalen Kongress fur Hygiene und Demographie].

Ross, R., 1909e. Malaria in Greece. Washington, DC: Government Printing Office. 697-710 pp.

Ross, R., 1910e. Report on the measures taken against malaria in the Lahore (Mian Mir) cantonment. Lancet 2: 1371-3.

Ross, R., 1910f. The fight against malaria. Standard (15 October), p. 4.

Ross, R. (+ 20 more collaborators), 1910. The prevention of malaria. John Murray, London (Sept., 1910) (2nd ed. June, 1911, with math. Addendum on the " Theory of Happenings". 711 pp.

Ross, R., 1911e. Lyra modulata {Lyra modulata bae Ronald Ross, Crysmas 1911) (Liverpool: [privately printed]).

Ross, R., 1913d. Patent acts and medical research. The Times (13 February), p. 13.

Ross, R., 1913e. Medical research. The Times (25 June), p. 4.

Ross, R. (1924) 'The transmission of human malaria', Nature, 113, p. 353.

Ross, R., 1914e. Cyprus as a health resort. Hospital 55: 611.

Ross, R., 1908a. Jl. Royal Army Medical Corps 10: 155-263.

Ross, R., 1893b. *ibid.* Indian Lancet.

Ross, R., 1893c. Nodulated and vacuolated corpuscles, and the solution of corpuscles mistaken foe parasites Indian Medical Records.

Ross, R. (1893f) 'Solution of corpuscles mistaken for parasites', IMR, 4, pp. 310-1.

Ross, R., 1894. The third element of the blood and the malaria parasite, and a list of natural appearances in the blood which have been mistaken for forms of the malaria parasite. Indian Medical Gazette.

Ross, R., 1895a. Parker's Memorial Prize Essay on malaria. Army Medical School, Netley. (unpublished).

Ross, R., 1895b. The crescent-sphere-flagella metamorphosis of the malaria parasite in the mosquito. Trans. S. Indian Branch Brit. Med. Assoc. 6: 334.

Ross, R., 1895c. The crescent-sphere-flagella metamorphosis of the malaria parasite in the mosquito. Transactions of South Indian Branch of British Medical association 6:334-350.

Ross, R., 1896a. Observations on the malarial parasite within the mosquito. Indian Medical Gazette

Ross, R., 1896b. Malaria parasites in Secunderabad. British Medical Journal (1st Feb 1896).

Ross, R., 1896c. Dr Manson's Mosquito-Malaria Theory. Indian Medical Gazette. (July 1896)

Ross, R. 1896d. Surg. Lt.-Colonel Lawri and the parasite of malaria. Indian Lancet. (Oct 1896)

Ross, R., 1897a. On some peculiar pigmented cells found in two mosquitoes fed on malaria blood. *British Medical Journal* 2: 1786-1788.

Ross, R., 1897b. Some experiments in the production of malaria fever by means of the mosquito. Indian Medical Gazette (Dec. 1896).

Ross, R., 1897c. On a condition necessary to the transformation of the malaria crescent. British Medical Journal (30 Jan 1897).

Ross, R., 1897d. Further observations on the transformation of crescents. South Indian Branch, Indian Medical Association (July, 1897).

Ross, R., 1897f. Observations on a condition necessary to the transformation of the malaria crescent. British Medical Journal 1: 251-255.

Ross, R., 1897g. Further observations on the transformation of crescents. South India Branch of British Medical Association.

Ross, R. 1897i. Notes on some cases of malaria. *Amoeba coli* and cercomonas. Indian Medical Gazette (Mar 1897).

Ross, R., 1897h. About a year and a half ago. Pioneer 1.

Ross, R., 1898a. Further observations on the transformation of crescents. Indian Medical Gazette (July, 1898).

Ross, R., 1898b. Preliminary report on the infection of birds with *Proteosoma* by the bites of mosquitoes. Government Press Calcutta (11 Oct, 1898); Indian Medical Gazette (Jan. 1899); British Medical Journal (18 Feb., 1899).

Ross, R., 1898c. Pigmented cells in mosquitoes. British Medical Journal (26 Feb 1898).

Ross, R., 1898d. Report on the cultivation of Proteosoma, Labbé, in grey mosquitoes, Calcutta. Reported in Indian Medical Gazette 23: 401-448.

Ross, R., 1898g. Cultivation of *Proteosoma* Labbe in grey mosquitoes. Office of the Superintendent of Government Printing, Calcutta, India.

Ross, R., 1898h. Pigmented cells in mosquitoes. British Medical Journal 1: 550-551.

Ross, R., 1898i. Report on the preliminary investigation into malaria in the Sigur Ghat, Ootacamund. Tv South Indian Branch of British Medical Association 7: 143.

Ross, R., 1898j. Report on the cultivation of *Proteosoma,* Labbe, in grey mosquitoes. Government Press, Calcutta, May 21st.Ross, R., 1899. Report to the Secretary to the Director General, Indian Medical Service, Calcutta.

Ross, R., 1898k. The role of the mosquito in the evolution of the malaria parasite. Lancet 2:489.

Ross, R., 1899r. Extermination of Malaria: *Ind. Med. Gaz.* 231-232 pp.

Ross, R., 1899a. Life history of the parasites of malaria. Nature (3 Aug. 1899).

Ross, R. 1899b. The cultivation of the malarial quartan parasite in Anopheles. *British Medical Journal* 2: 608.

Ross, R., 1899c. Report to the Secretary to the Director General, Indian Medical Service, Calcutta.

Ross, R., 1899d. Life history of the parasites of malaria. Nature.

Ross, R., 1899f. Mosquitoes and malaria. *British Medical Journal.*

Ross, R., 1899g. Infection of birds with *Proteosoma* by the bites of mosquitoes. Indian Medical Gazette 33:401-448.

Ross, R., 1899h. Report on the nature of Kala-azar. Office of the Superintendent of Government Printing, Calcutta, India.

Ross, R., 1899i. Du Rôle des mosquitoes dans le Paludisme . Acad. De médicine, Paris (24 Jan) and Ann de l' Inst. Pasteur, 1899, p. 136.

Ross, R., 1899k. Letter to Government of India on malaria prevention by means of mosquito reduction . Indian Medical Gazette (July, 1899)

Ross, R., 1899l. The possibility of extirpating malaria from certain localities by a new method. Inaugural Lecture at University College, Liverpool. British Medical Journal. (1st July 1899).

Ross, R., 1899m. The Malaria Expedition to Sierra Leone. *British Medical Journal* 2: 675-676.

Ross, R., 1899n. The Malaria Expedition to Sierra Leone: Mosquito-Borne Fever at Wilberforce. *British Medical Journal* 2: 746.

Ross, R., 1899o. The Malaria Expedition to Sierra Leone: Anopheles and its Habits Malarious Foci Localised. *British Medical Journal* 2: 869-871.

Ross, R., 1899p. The Malaria Expedition to Sierra Leone: Habits of Anopheles Continued.-Possibility of Extirpation.-Explanation of the Old Laws of Malaria. *British Medical Journal* 2: 1033-1035.

Ross, R., (Anon.) 1899q. *Instructions for the prevention of malarial fever.* University Press. Liverpool, 14 pp.

Ross, R., H.E. Annett and J.E. Dutton, 1899. *Report on malaria expedition to West Africa.* Liverpool School of Tropical Medicine, Mem. 2.

Ross, R., 1915-16. The solution of equations by operative division. Part I, 10, pp. 218-236; Part II, 10, pp. 393-411; Part III , 10, pp. 573-600.

Ross, R., 1900a. Malaria and mosquitoes. Lecture to Royal Institution, 2nd March 1900. Nature (29 March) revve Scientifique (23 June, 1900).

Ross, R. 1900b. Malarial fever. *The Medical Annual Journal*, Wright & Co., Bristol.

Ross, R., 1900c. Le scoperte del Prof. Grassi sulla malaria: Two papers policlinico (Nov. 1900 and 1 May 1900)

Ross, R., 1900. Letters from Rome on the new discoveries on malaria. Liverpool [privately printed]).

Ross, R., H.E. Annett and E.E. Austen, 1900a. *Report of the malaria expedition to West Africa, 1899.* Liverpool School of Tropical Medicine, Memoirs II.

Ross, R., H.E. Annett and E.E, Auston, 1900b. Report of the malaria expedition of the Liverpool School of Tropical Medicine and Medical Parasitology. Liverpool: University Press of Liverpool. 1-71 p.

Ross, R. and R. Fielding-onld, 1900. Diagrams illustrating the life history of the parasites of malaria. *Quarter Journal of Microscopic Science*, No. 171 (July, 1900)

Ross, R., 1901a. First Progress Report of the Campaign against mosquitoes in Sierra Leone. Universal Press, Liverpool (15 Oct., 1901)

Ross, R., 1901b. *The work of the Liverpool School of Tropical Medicine in West Africa.* Africa Section, Chamber of Commerce, Liverpool (21 Oct, 1901) – a pamphlet.

Ross, R., 1902a. *Mosquito brigades and how to organize them.* New York, Longmans, Green & Co.

Ross, R., 1902b. *Malarial fever; its cause, prevention and treatment, containing full details for the use of travelers, sportsmen, soldiers and residents in malarious places.* London, Green.

Ross, R., 1902c. *Malaria fever: its cause, prevention and treatment.* University Press, Liverpool (Feb., 1902), pp. 68.

Ross, R., 1902d. Article on malaria disease. Quain's Dictionary of Medicine, London (21 Columns with two plates)

Ross, R. (1902g) 'Sanitary affairs in West Africa', in E.D. Morel, Affairs of West Africa (London: Heinemann) pp. 151-69.

Ross, R. (1902h) 'Arsenic in the hair of patients from Penang. *British Medical Journal* 1: 329-30.

Ross, R., 1903b. Note on the bodies recently discovered by Leishman and Donovan. British Medical Journal 2: 1261-1262.

Ross, R., 1903d. *Report on malaria at Ismailia and Suez. Memoir.* Liverpool School of Tropical Medicine. (Jan., 1903).

Ross, R., 1903f. Papers on the thick-film process for malaria diagnosis. Lancet (10th Jan., 1903). *J. Trop. Med.* (2nd Feb, 1903).

Ross, R., 1903g. Papers on the thick-film process for malaria diagnosis. *Lancet* (10th Jan., 1903). Thompson Yates, Labor. Reports 5 (I).

Ross, R., 1903h. On the new discovered Leishman-Donovan parasites. Brit. Med. J. 14/21 and 28th Nov., 1903.

Ross, R., 1903i. On the new discovered Leishman-Donovan parasites. Thompson Yates Labson Reports, Vol. 5 (2).

Ross, R., 1903j. Malaria in India and the Colonies. *Journ. R. Colonial Institute*. (Dec, 1903).

Ross, R., 1903k. An improved method for the microscopical diagnosis of intermittent fever. The Lancet 161: 86.

Ross, R., 1903l. An improved method for the microscopical diagnosis of Ross, R., 1904. *J. Trop. Med. & Hyg*., p. 75.

Ross, R., 1903m. *The thick film process for the detection of organisms in the blood.* Thompson Yates & Johnson Laboratory Report 1 (1): 117-119.

Boyce, R.W., C.S. Sherrington and R. Ross, 1903. The history of the discovery of Trypanosomiasis in man. *Lancet* (21st Feb., 1913).

Ross, R., 1904a. An instrument for obtaining continuous clinical .... charts. *Lancet* (27th Feb., 1904).

Ross, R., 1904b. The anti-malaria experiment at Mian Mir. *British Medical Journal*: 1-50. (17 Sep., 1904).

Ross, R., 1904c. The logical basis of the sanitary policy of mosquito reduction. *Proceedings of the Congress of Arts and Sciences*, St Louis, USA 6: 89.

Ross, R., 1904d. The logical basis of the sanitary policy of mosquito reduction. *British Medical Journal* 1: 1025-1029. (13 May, 1905).

Ross, R., 1904f. *Researches on malaria.* (Nobel Medical Prize Lecture for 1902). P.A. Norstedt & Söner, Stockholm, 88 pp.

Ross, R., 1903a. The extirpation of Culex at Ismailia. *Lancet* 2: 186.

Ross, R., 1905a. An address on the logical basis of the sanitary policy of mosquito Reduction. *BMJ* 1: 1025-9.

Ross, R., 1905b. Verb-functions. *Proceedings of the Royal Irish Academy* 25, section A, 31-76.

Ross, R., 1905d. The progress of Tropical Medicine: A lecture delivered before Princess Christian at St. George's Hall, Liverpool (12th June, 1905). Journal of African Society.

Ross, R., 1906d. Notes on the parasites of mosquitoes found in India between 1895 and 1899. Journal of Hygiene 6: 101-109.

Ross, R., 1906a. Parasites of mosquitoes found in India between 1895 and 1899. J Hyg., Cambodia (April, 1906).

Ross, R., 1907b. The prevention of malaria in British possessions, Egypt, and parts of America. The Lancet 2: 879-887. (28th Sep).

Ross, R., 1907c. The public prophylaxis of malaria. Allbutt and Rollester's System of Medicine 2(2).

Ross, R., 1907d. Malaria campaign. The prevention of malaria in British possessions. [British uber den XIVen Internationalen Kongress for Hygiene and Demographie].

Ross, R., 1907f. Introduction. (*in* W.H.S. Jones. Malaria, a neglected factor in the history of Greece and Rome. Cambridge: Macmillan & Bowes, pp. 1-14.

Ross, R., J.E. Salvin-Moore and C.E. walker, 1907a. A new microscopical diagnostic method., etc. Lancet. (27 July, 1907).

Ross, R., J.E. Salvin-Moore and C.E. walker, 1907b. On the existence of centrosomes .... in red blood corpuscles. Pathol. Soc. London 58 (1).

Ross, R., 1908b. Report on the prevention of malaria in Mauritius. London, J. & A. Churchill, 1909, 202 pp.

Ross, R., 1908c. Report on the prevention of malaria in Mauritius. London: (privately printed), 202 pp.

Ross, R., 1908d. Report on the Prevention of Malaria in Mauritius. New York: E. P. Dutton & amp. Company, 202 pp.

Ross, R., 1909a. The campaign against malaria. Royal Institution, London (7 May, 1909).

Ross, R., 1909b.The practice of malaria prevention. Transactions Bomabay Medical Congress.

Ross, R., 1909c. The practice of malaria prevention. Trans. Bombay Medical Congress.

Ross, R., 1909d. Report on the Prevention of Malaria in Mauritius. London: J. & A. Churchill.

Ross, R., 1909. Mosquitoes and malaria: the campaign that failed. Lancet 1: 1074-5.

Ross, R., 1910a. The prevention in malaria. London, John Murray.

Ross, R., 1910b. The Prevention of Malaria. New York, F.P. Dutton & Co., 661 pp.

Ross, R., 1910c. The measures taken against malaria in the Lahore (Mean Mir) Cantonment. Lancet (5 Nov., 1910).

Ross, R., 1910d. The more exact study of parasitic diseases. Transactions of the Society of Tropical Medicine and Hygiene, 3, pp. 383-90.

Ross, R., 1910g. A summary of facts regarding malaria suitable for public instruction. London: John Murray.

Ross, R. and D. Thomson, 1910a. Some enumerative studies on malarial fever. Annals of Tropical Medicine & Parasitology 4:267-306.

Ross, R. and D. Thomson, 1910b. Some enumerative studies on malaria. Annals of Tropical Medicine & Parasitology 4: 267.

Ross, R. and D. Thomson, 1910. Some enumerative studies on malaria fever. *Proceedings of the Royal Society of London.* Series B, Containing Papers of a Biological Character 83: 159-173.

Ross, R., 1911a. Some quantitative studies in epidemiology. Nature 87: 466-467. (5 Oct., 1911)

Ross, R. and W. Scott, 1911b. Tables of statistical errors. Ann. Trop. Med., Liverpool (Dec 1911).

Ross, R. (& Contributors), 1911c. The prevention of malaria (second edition), London.

Ross, R., 1911d. The prevention of malaria: a review reviewed. IMG, 46: 154-5.

Ross, R. & D. Thomson, 1911. Pseudo-relapses in cases of malarial fever during continuous quinine treatment. Annals of Tropical Medicine & Parasitology 5: 409.

Ross, R. & D. Thomson, 1912. A case of malarial fever showing a true parasitic relapse during vigorous and continuous quinine treatment. Annals of Tropical Medicine & Parasitology 5: 539.

Ross, R., S. Thomson, J.G. Thomson and GC.E. Simpson. 1912. A series of enumerative methods and effects of cold-chamber and other treatment in malaria and trypanosomiasis. Proc. R. Soc., Soc. Trop. Med., and Ann. Trop.Med., Liverpool (June 1910 to 1912).

Ross, R., 1913a. Medical science and the tropics. Royal Colonial Institutes (14 Jan., 1913).

Ross, R., 1913f. Tropical medicine. Appeal from Sir R. Ross. The Times (16 August), p. 4.

Ross, R., 1914c. The reward of research: how to apply for it. BMJ 1: 941-2.

Ross, R., 1914d. Correspondence concerning a petition. London [privately printed]).

Ross, R., 1914f. Report on the prevention of malaria in Cyprus. Presented to both Houses of Parliament by Command of His Majesty. Printed under the Authority of His Majesty's Stationery Office by Darling and Son, Ltd., Bacon Street, E., London.

Ross, R., 1914. Intramuscular injections of quinine. (Correspondence) Journal of Trpical Medicine & Hygiene 27: 286.

Ross, R., S.R. Christophers & E.L. Perry, 1914. The spleen rate in London School children. Indian Journal of Medical Research 1:3.

Ross, R., 1915. Some a *priori* pathometric equations. *British Medical Journal* 1: 546-547.

Ross, R., 1916. An application of the theory of probabilities to the study of a priori pathometry. Pt. I. Proc. R, Soc., A. Vol. 92(2): 204-230.

Ross, R. and Hilda P. Hudson, 1916. An application of the theory of probabilities to the study of a priori pathometry. Pt. I. Proc. R, Soc., A. Vol. 92; Pt. II and III with Hilda P. Hindson, *ibid.* A. Vol. 93, Year 1917.

Ross, R., 1913b. Medical science and the tropics. United Empire 4: 120-35.

Ross, R. and Hilda P. Hudson, 1917. An Application of the Theory of Probabilities to the Study of a priori Pathometry. Part II. Proceedings of the Royal Society of London Series A. Mathematical Physical and Engineering Sciences 93: 212-225.

Ross, R. and H.P. Hudson, 1917. An Application of the Theory of Probabilities to the Study of a priori Pathometry. Part III. Proceedings of the Royal Society of London Series a-Mathematical Physical and Engineering Sciences 93: 225-240.

Ross, R. (Editor), 1918b. Malaria in Macedonia (by Armand-Dellile, P. et al.). University of London.

Ross, R. (editor), 1919. Observations on malaria by medical offices of the Army and others. War Office Publication, H.M. Stationary Office, London (Dec., 1919). Ten Reports, 342 pp.

Ross, R., 1921. The principle of repeated medications for curing infections.. *Brit. Med. J.* 2: 1-4. (2 July, 1921).

Ross, R., 1906f. The Lancet (Nov 17).

Ross, R., 1923b. Hygiene for Indian scholars. (Bombay: K. & J. Cooper).

Ross, R., 1925b. The treatment of malaria in Britain. The Practice (Nov., 1925).

Ross, R., 1926a. Malaria control in Ceylon plantations. Journal of Tropical Medicine & Hygiene 29: 150-155.

Ross, R., 1926b. Report on malaria control in Ceylon plantation. A report of the Ceylon Association, London (Apr, 1926)

Ross, R., 1927a. Malaria control in Malaya and Assam. A report on a visit of inspection, 1926-27. (Indian Tea association, London, 21 Mincing Lane, E.C. 3).

Ross, R. (1928c) Studies on malaria (London: John Murray)

Ross, R., 1928a. Mosquito Control: General or special. The Practitioner, London (Oct., 1928).

Ross, R., 1928b. The progress of malaria control. New Health (Dec., 1928).

Ross, R. (1929a) Letters from Rome on certain discoveries in malaria, being a series of letters written from Rome by T. Edmonston Charles (London: [privately printed]).

Ross, R., 1929a. The amount of malaria depends on the numbers of carriers. London, John Murray.

Ross, R., 1929b.Constructive epidemiology. British Medical Journal 1: 673-674.

Ross, R. (1930a) Memories of Sir Patrick Manson (London: [privately printed]).

Ross, R. (1930b) The solution of equations by iteration (London: [privately printed]).

Ross, R and H.P. Hudson, 1931. A priori pathometry. Harrison and Sons, Ltd, London: 75.

Ross, R., 1907a. The Lancet (Sep 28).

Ross, R., 1930. Journal of Tropical Medicine & Hygiene (Apr 28, 1930).

Ross, R., 1903c. A new parasite of man. *TY & J. Lab. Report* 5: 79-82.

Ross, R., 1907g. Malaria in Greece. University Review 5: 101-19.

Ross, R., 1998f. Preliminary report on the infection of birds with Proteosoma by the bites of mosquitoes. Calcutta Government Press.

Ross, R. and Hilda P. Hudson, 1916c. Pts. II & III. Vol. 93, Year 1917. *Brit. Med. J.* (27 Mar., 1915).

Ross, R., 1916-17. Petition. Science Progress 11: 133.

Ross, R., 1928c. Studies on malaria. London, John Murray. 196 pp.

Ross, R., 1914a. Tropical medicine - a crisis. BMJ 1: 319-21.

Ross, R., 1905c. Yellow fever in New Orleans. BMJ 2: 414.

Ross, R., 1923a. Memoirs. John Murray. London. 523 pp.

Ross, R., 1925a. Annals d'Ig pag, 532.

Ross, R., 1913g. The double victory at Panama. Nation 13: 566-7.

Ross, R., 1914b. The penguin people. Nation 15: 664-6.

Ross, R., 1922. The work of Sir Patrick Manson. BMJ 1: 698.

Ross, R., 1913c. Malaria prevention in Greece. BMJ 1: 1186.

Ross, R., 1906c. Malaria in Greece. University Review (May, 1907).

Ross, R., 1929c. Africa world, rec. Journ A.M.A. 1929, Image eRiv di Mal 1929, maggie, page. 335.

Ross, R., 1927b. 'Is discovery inevitable?', Sci. Prog., 1927, 22: 130.

Stephens, J.W.W., S.P. James, R. Ross, E.P., Sewell H., Strachan *et al.*, 1904. Discussion on the prophylaxis of malaria. *British Medical Journal* 2: 629-642.

W.H.S. Jones, R. Ross and G.G. Ellett, 1907. Malaria – a neglected factor in the history of Greece and Rome. MacMillan & Bowes, London, pp. 108.

Annex 2. **Non-malariological Bibliography of Dr Ronald Ross**

Ross, R. (1928) Fables and satires (London: [privately printed]).

Ross, R. (1928) Poems (London: Elkin, Mathews & Marrot).

Ross, R. (1929) Solid space algebra: the systems of Hamilton and Grassmann combined (principles only - seventh draft, 1918) (London: [privately printed]).

Ross, R. (1931-2) 'Doktorology', Science Progress, 26, pp. 133-4.

Ross, R. (1931a) In exile (London: [privately printed])

Ross, R. (1931b) Lyra modulata (London: [privately printed]).

Ross, R. (1931c) Circles by explicit operations (London: [privately printed]).

Ross, R. (1932) The child of ocean, rev. edn (London: George Allen & Unwin).

Ross, R. and Boyce, R.W. (1902) 'The discovery of the human trypanosome', Hospital, 33, p. 135.

Ross, R. and Milne, 1904. Tick fever. British Medical Journal 26(11): 1453.

Ross, R. and Reynolds, E.S. (1901) 'A case of beri-beri (?) possibly due to arsenic poisoning', BMJ, ii, pp. 979-80.

Ross, R. and Thomson, D. (1915-16) 'Studies on Egyptian sand amoebae', Proceedings of the Royal Society of Medicine, 9(2), pp. 33-48.

Ross, R. *et al.* (1907) 'India and Mr Haffkine', The Times, 1 June, p. 8

Ross, R., 1892a. Fever with intestinal lesion. South Indian Branch of British Medical Association.

Ross, R., 1892b. Cases of febricula with abdominal tenderness. Indian Medical Gazette

Ross, R., 1892c. Enteroseptic fevers. Indian Medical Gazette.

Ross, R., 1892d. A study of Indian fevers. Indian Medical Gazette.

Ross, R., 1905a. An address on the logical basis of the sanitary policy of mosquito Reduction. BMJ 1: 1025-9.

Ross, R., 1905b. Verb-functions. Proceedings of the Royal Irish Academy 25, section A, 31-76.

Ross, R., 1905c. Yellow fever in New Orleans. BMJ 2: 414.

Ross, R., 1906a. A physiological diacritical vowel system. JOO 23: 23-8.

Ross, R., 1906bc. Some remarks. JOO 23: 47-9.

Ross, R., 1906c. An experimental measure for the Iliad? JOO 23: 90-1.

Ross, R., 1907a. India and Mr Haffkine. The Times, 15 March, p. 4.

Ross, R., 1907b. India and Mr Haffkine. The Times, 13 April, p. 8.

Ross, R., 1907c. The inoculation accident at Mulkowal. Nature 75: 486-7.

Ross, R., 1910. Philosophies. London: John Murray.

Ross, R., 1911. Lyra Modulata. the University of Wisconsin - Madison [privately printed], 9 pp.

Ross, R., 1913. Man and woman. Nation 14: 93-4.

Ross, R., 1916. A lecture on the treatment of dysentery...'. Lancet 1: 1-7.

Ross, R., 1916-17. Sister of science. Science Progress 11: 441-6.

Ross, R., 1917a. Conflagration. Poetry Review 8: 75-7.

Ross, R., 1917b. Death of peace. Poetry Review 8: 1-4.

Ross, R., 1918-19. Isosceles trigonometry: suggested addition to the trigonometrical ratios. Science Progress 13: 485-6.

Ross, R., 1918-19. Operative algebra: operative involution. Science Progress 13: 288-98.

Ross, R., 1918a. A national house of poetry. Poetry Review 9: 57-9.

Ross, R., 1918b. Lines addressed to the Poetry Society by the President. Poetry Review 9: 69-70.

Ross, R., 1919-20. A great default. Science Progress 14: 664-8.

Ross, R., 1920. The revels of Orsera: a mediaeval romance. London: John Murray.

Annex 3. **New Species of Parasites Discovered/ Described by Dr Ronald Ross**

The following are the parasites discovered by Ronald Ross in mosquitoes:

1. A nematode worm
2. *Gregarina culicis* (probably several species)
3. *Coccidium culicis*
4. *Octosporidium culicis*
5. Certain flagellate organisms

Annex 4. **Ph.D./M.Sc. Theses about Dr Ronald Ross**

1. Beaumonte, B.E., 1974. Sir Ronald Ross: a bio-bibliography. M.A. Thesis, Queen's University, Belfast.

Annex 5. **Post-retirement from British Indian Medical Services Important Positions/ Assignments/ Consultancies Rendered by Dr Ronald Ross**

1. In 1899 he joined the Liverpool School of Tropical Medicine under the direction of Sir Alfred Jones.
2. In 1902 he was appointed a Companion of the Most Honourable Order of Bath by His Majesty the King of Great Britain.
3. 1911 (to 1913), Vice-President of Royal Society from.
4. In 1911 he was elevated to the rank of Knight Commander of the same Order. In Belgium, he was made an Officer in the Order of Leopold II.
5. In 1902 a movement was set on foot to commemorate the valuable services rendered to the School of Tropical Medicine by its originator and Chairman, Sir Alfred Jones, by founding a Chair of Tropical Medicine in University College to be connected with the School. The movement was met with enthusiastic support, and an amount of money was quickly collected sufficient to found «Sir Alfred Jones' Chair of Tropical Medicine». Ross was appointed to the Professorship in 1902.
6. He retained the Chair of Tropical Medicine until 1912.
7. 1912 - 1917, when he left Liverpool, and was appointed Physician for Tropical Diseases at Kings College Hospital, London, a post which he held together with the Chair of Tropical Sanitation in Liverpool.
8. 1917, he was appointed Consultant in Malariology to the War Office.
9. 1918, Knight Commander, St. Michael and St. George.
10. He was later appointed Consultant in Malaria to the Ministry of Pensions.
11. In 1926 he assumed the post of Director in Chief of the Ross Institute and Hospital of Tropical Diseases and Hygiene, which had been created by admirers of his work, and he remained in this position until his death.
12. He was also a President of the Society of Tropical Medicine.

Annex 6. **Laurels, Distinctions, Awards, Prizes and Recognitions – *The end of an Era with an Aura***

HONOURS AND AWARDS

Ronald Ross was exceptional in all spheres of life. This has been evident from the awards and honours that was bestowed upon him. A few have been enlisted according to the year in which Ross received them.

1895 : Ross won the Parkes Gold Medal of the Army Medical School, Netley for an essay that he wrote (Bynum and Overy,1998).

1901 : A) Ross received the honour as a Fellow of Royal College of Surgeons, England on 11th April,1901. B) He was also elected FRS on 7th June,1901 (Bynum and Overy,1998). C) Ross also received the Silver Medal from the Society of Arts, London on 15th June,1901.

1902 : 1902 was an auspicious year in Ross' life. He received back to back three awards during this period. A) One being the honour of the Companion of Bath on 26th June, 1902. B) The other was the Cameron Prize, Edinburgh University, July 26 the very same year. C) The greatest achievement of Ross was being awarded the Nobel Prize for medicine on 10th December, 1902 (Nye and Gibson, 1997; cited by Bynum and Overy, 1998) which created a landmark in the history of medicine.

1903 : The stay of Ross in Bengal also became worthwhile when he received the Barclay Bronze Medal on 20th May,1903 given by the Asiatic Society of Bengal.

1904 : Ross was also given the Sc.D from the Trinity College, Dublin in the year 1904 on the 20th of December.

1906 : A) Ross was titled Officer de l'Ordre de Leopold II on 31st August,1906 at Belgium. B) He also got the LL.D in Aberdeen on 26th September the same year.

1909 : A) In the year 1909, on the 9th of July Ross was honoured as D.Sc., Leeds. B) Interestingly he also received the Royal Gold from the Royal Society on November 18,1909.

1910 : The Karolinska Institute, Stockholm gave Ross the M.D on December 13, 1910.

1911-1913 : The start of 1911 saw a golden period in Ross' life starting from being coronated as A) The Knight Commander of the Order of Bath in the Coronation Honours of George V- an extremely prestigious title (Bynum and Overy, 1998). B) Between the years 1911-1913 Ronald Ross played a major role residing as the Vice-President of the Royal Society and in between the time frame C) Ross was offered M.D., Athens on 12th April,1912 D) He also received the honour of Officer de I'Instruction Publique, France on November 1, 1913.

1914 : In the subsequent year, Ross received the Bisset Hawkins Gold Medal, by the Royal College of Physicians in London in 19th October, 1914.

1915 : The Freeman by the Society of Apothecaries on March 9, 1915 at London was given to Ross.

1918 : Apart from being the Knight Commander of the Order of Bath in the year 1911, on 3rd June, 1918 Ross was also appointed the Knight Commander by St. Michael and St. George.

1922 : Ronald Ross was included in the Athaenum Club (Rule II) on 11th April,1922 and as a member of the R. Soc. of Literature in London on 15th May, 1922.

Ross had commendable achievements in a span of a few months from everywhere in the world.

Annex 7. **Significant Prizes and Awards Won by Dr Ronald Ross**

1. Ronald Ross was awarded the Nobel Prize for Physiology or Medicine in 1902 "for his work on malaria, by which he has shown how it enters the organism and thereby has laid the foundation for successful research on this disease and methods of combating it".
2. 20 August is celebrated by London School of Hygiene & Tropical Medicine as *World Mosquito Day* to commemorate Ross' discovery in 1897.
3. Ross' name, along with 22 other pioneers of public health and tropical medicine, appears on the School's Frieze.
4. The papers of Sir Ronald Ross are now preserved by the London School of Hygiene and Tropical Medicine[3][32] and the Royal College of Physicians and Surgeons of Glasgow.
5. He was elected a Fellow of the Royal Society (FRS) in 1901.
6. Fellow, Royal College of Surgeons in the same year.
7. He was appointed Vice-President of the Royal Society from 1911 to 1913. In 1902 he was appointed a Companion of the Most Honourable Order of the Bath by King Edward VII. In 1911 he was promoted to the rank of Knight Commander of the same Order. He was also decorated with the title Officer of the Order of Leopold II of Belgium.
8. Ross received honorary membership of learned societies of most countries in Europe, and elsewhere. He got an honorary M.D. degree in Stockholm in 1910 at the centenary celebration of the Caroline Institute.
9. His 1923 autobiography *Memoirs* was awarded that year's James Tait Black Memorial Prize.
10. In India, Ross is remembered with great respect as a result of his work on malaria, the deadly epidemic which used to claim thousands of lives every year.
    (i) There are roads named after him in many Indian towns and cities. In Calcutta the road, known as Hospital road, linking Presidency General Hospital with Kidderpore Road has been renamed after him as Sir Ronald Ross Sarani.

(ii) In his memory, the regional infectious disease hospital at Hyderabad was named Sir Ronald Ross Institute of Tropical and Communicable Diseases.

(iii) Sir Ronald Ross Institute of Parasitology was established in memory of Ronald Ross in Hyderabad, under Osmania University. This building where he worked and actually discovered the malarial parasite, located in Secunderabad near the Begumpet Airport, is declared a heritage site, and

(iv) The road leading up to the building is named Sir Ronald Ross Road.

(v) In Ludhiana, Christian Medical College has named its hostel as "Ross Hostel". The young medics often refer to themselves as "Rossians".

11. The University of Surrey, UK, has named a road after him in its Manor Park Residences.
12. Ronald Ross primary school near Wimbledon Common is named after him. The school's coat of arms includes a mosquito in one quarter.
13. In 2010 the University of Liverpool has named its new biological science building "The Ronald Ross Building" in his honour. His grandson David Ross inaugurated it. The building is home to the university's facility for the Institute of Infection and Global Health.

Annex 8. **Fellowship and Honorary Memberships of Scientific Societies Conferred on Dr Ronald Ross**

Ross was given his due respect by granting him honorary memberships throughout his lifetime.

1901	:	Ross was elected a Fellow of the Royal College of Surgeons of England.
1901	:	a Fellow of the Royal Society.
1901	:	Ross became a member of the Medical Society of Netherlands, Batavia on 20th March 1901.
1902	:	Soon after, within a span of a year he became the member of Soc. de. Med. De Gand on the 14th of January 1902.
1903-1903	:	was a remarkable year in Ross' life with a handful of memberships being offered to him. One of the many firsts being (A) The member of Soc. Medicorum Svecana in the month of January,1903. This was followed by (B) The R. Physiographie Soc. Lund on 11th February the same year. (C) He also became the member of the Manilla Med. Soc. on March 1,1903. (D) Ross got embraced into the American Soc. Tropical Medicine on 19th June,1903. (E) Ross was also included in the Soc. Med. et Hyg. tropicales,Paris on 11th December 1903.
1904	:	Ross got appended into the Lancashire and Chesire Entomolog. Soc. on the 18th of January,1904 and Acad.de Med. Paris,19th July,1904.
1906	:	The Malaria Soc. of Greece paved the way for Ross' membership here on 12th July 1906.
1907	:	R. Accad. di Med., Torino (Corr. estero) in the year November 9,1907 included Ross in their Society.
1908	:	The year 1908 also proved to be a significant event in Ross' life with three consecutive membership like (A) The Soc. de Pathologie exotique in Paris, 22nd January 1908. (B) The Coll. of Physicians, Philadelphia on 5th February,1908. (C) The Soc.

de Med. Trop.,Cuba on April 25th,1908 and (D) The Hon. President de Med. Soc. Mauritius on December 4,1908.

1909-1911 : Ross inclusively became the President of Soc. Trop. Med. London, from a period of 18th June 1909-1911.

1912 : With time Ross became more and more respected and accepted in the fraternity of memberships and A) He became a member of the Russian Malaria Commission, on February 2,1912. B) He also became a member of the Imp. Military Acad., St. Petersburg, February 2 in the year 1912.

1913 : Ross progressively became a part of A) the Haverian Soc. of London , in October 1913 and B) the Inst. of Hygiene, London in December 1,1913.

1919 : Ross became A) An integral part of the Soc. of Sciences, Upsala on 5th April,1919 and B) The Soc. Belge de Biologie, on December 6,1919.

1920 : Ross was a member of the Ecole de Med. trop. Bruxelles on December 30,1920.

1921 : Ross became the member of both R. Soc. of Edinburgh (4 July) and Acad. de Med. Paris (27 December) in the year 1921.

Annex 9. **Literary Works of Ronald Ross**

Science and Literature are orchestrated and the understanding of one could lead to the expression of another. Ross was a prolific writer. He habitually wrote poems on most of the important events in his life. His poetic works gained him wide acclaim and they reflect his medical service, travelogue, philosophical and scientific thoughts. He carefully saved virtually everything about himself: correspondence, telegrams, newspaper cuttings, drafts of published and unpublished material, and all manner of ephemera, amounting close to 30, 000 items!

1883 : Edgar, or the New Pygmalion ; and The Judgement of Tithonus: Higginbotham, Madras having 151 pages and two dramas in verse.

1889 : The Child of Ocean; a Romance: Remington, London consisting of 344 pages.

1890 : The Deformed Transformed which was privately printed by Spectator press in Bangalore and had 128 pages. This was later published by Chapman & Hall in London in the year 1892 having 157 pages.

1896 : The Spirit of Storm; a Romance: Methuen, London with 352 pages.

1906 : In Exile was also privately printed by Philip, Son and Nephew in Liverpool consisting of 84 pages. Verses written were 1890-7. This too was later on published with the omission of some stanzas in No. 114.

1906 : Collection of poems "Selected Poems" (1928) and In Exile (1931).

1907 : Fables: University Press in Liverpool published this with 51 pages. Verses written were 1881-96 and Twenty fables were written in those verse.

1910 : Philosophies: by Murray . Subsequent versions were also released in 1911 having 56 pages and verses written were 1881-99.

1911 : Lyra Modulata was also printed privately by C. Tinling in Liverpool,quarto with 7 pages having ten poems in phonetic spelling.

1912 : The Setting Sun : Murray, 64 pages. It had a satire in verse.

1913 : (A) The Indian Shepherds; a Fragment : English Review printed it in the month of September having 7 pages. Interestingly, its verses commenced in 1898. (B) Otho: The Nation was printed on

September 27 the same year. (C) The Triumph: The Nation on 13th December. These were short dramas in verse reprinted in 1921.

1916 : (A) Shakespeare: Book of Homage printed by University Press, Oxford. (B) Prayer: Science Progress in London, in the month of July. (C) Single sonnets like The Fall of the Zeppelin: the Nation was also printed on 14th October.

1917 : (A) Verses on the War : Poetry Review, London during the months of January-March. (B) A treasury of War Poetry: Cambridge, U.S.

1919 : (A) Another League of Nations; a Fable: By Morning Post published it on the February 14. (B) Psychologies: Murray published it which consisted of 69 pages. It was a compilation of 5 short dramas in verse.

1920 : The Revels of Orsera; a Medieval Romance: Murray. It had 393 pages.

No. 110 eventually became a Novel and was written in the year 1895.

Annex 10. **Dr Ross - The Mathematician**

Ronald Ross had a predisposition towards the subject of mathematics which reflects itself in several mathematical papers by him. This rightly justifies why we could call Ross a multi-faceted scientist, a polymath, whose expertise in different disciplines resonates in many directions. The following are the mathematical papers written by Ross:

1. 1901 :- The Algebra of Space being a brief description of a System of Geometrical Algebra placed on an Arithmetical Basis : G.Philip & Son,32 Fleet Street,London. This paper had 31 pages,quarto and combined the systems of Hamilton and Grassmann.
2. 1905 :- Verb-functions : Proc. R. Irish Acad., Vol XXV, A.3, April. This paper suggests a notation for rendering operations explicitly apart from their subjects; studies the elementary algebra of the subject,namely operative multiplication and division; employs the latter to solve algebraic equations of any degree and some differential equations; and shows how the notation increases the power of our mathematical algorithm.
3. 1908-1909 :- A Method of Solving Algebric Equations : it had two papers published by Nature on the 29th of October, 1908 and 4th of February, 1909. This paper is an anthology that rediscovers the method of "iteration" originally found in an elementary way by Michael Dary, a friend of Newton's, and forgotten in Britain except in the form used by Newton.
4. 1915-1916 :- The Solution of Equations by Operative Division: Science Progress,Vol. X, published in October 1915 and January and April 1916. This paper elaborates and combines the results of the three previous papers, General formulae and eight figures.
5. 1918 :- Operative algebra; Operative Involution: Science Progress, Vol. XIII, in January. This keeps its continuity with the theme of Verb-Functions.
6. 1919 :- Isoscles Trigonometry: Science Progress, Vol. Xiii, in January. This consolidates the trigonometrical ratios based on the isosceles triangle.
7. 1921 :- The Operative Roots of the Circle-function: Science Progress, Vol. XVI, in July which potrays the De Moivre's Theorem concerning imaginary numbers is a particular case bof a more general theorem in operative involution concerning the operative ratios of chords of a circle - which degenerates to De Moivres' Theorem when the diameter of the circle is zero.

Annex 11. **Books by Dr Ronald Ross**

1. Ross, R. (1889) The child of ocean (London: Remington).
2. Ross, R. (1890) The deformed transformed (Bangalore: Spectator Press).
3. Ross, R. (1892) The deformed transformed (London: Chapman & Hall).
4. Ross, R. (1896) The spirit of storm (London: Methuen).
5. Ross, R. 1901. The algebra of space. London: Philip.
6. Ross, R. 1910. Philosophies. London: John Murray.
7. Ross, R. 1916-17b. Homer, Dante, Shakespeare and Cervantes. Science Progress 11: 137-40.
8. Ross, R. 1916a. Shakespeare. *in* I. Gollancz (ed.), A book of homage to Shakespeare (Oxford University Press).
9. Ross, R. 1916c. The fall of the Zeppelin. Nation 20: 84.
10. Ross, R. 1919. Psychologies. London: John Murray.
11. Ross, R. 2010 [Reprint]. Fables. Kessinger Publishing, LLC.

Annex 12. **Unpublished Reports by Dr Ronald Ross**

- Report on Cholera, General Sanitation, and the Sanitary Department and Regulations, in the C. & M. Station of Bangalore (1896)
- Report on the Cultivation of Proteosoma Labbé, in Grey Mosquitoes (1898). Digitised version available from National Library of Scotland.
- Report on the Nature of Kala-azar (1899). Digitised version available from National Library of Scotland.
- Malarial Fever: Its Cause, Prevention and Treatment; Containing Full Details for the Use of Travelers, Sportsmen, Soldiers, and Residents in Malarious Places (1902)
- First Progress Report of the Campaign Against Mosquitoes in Sierra Leone (with Charles Wilberforce Daniels) (1902)
- Notes on the Parasites of Mosquitoes Found in India Between 1895 and 1899
- Hygiene for Indian Scholars
- Note on the Bodies Recently Described by Leishman and Donovan (1903)
- Further Notes on Leishman's Bodies (1903)
- Report on Malaria at Ismailia and Suez (1903)
- Leishmania donovani Found in Kala-azar (1904)
- Researches on Malaria (1905)
- Note on a Flagellate Parasite Found in Culex fatigans (1906)
- Malaria in Greece (1909)
- Missionaries and the Campaign Against Malaria (1910)
- A Case of Sleeping Sickness Studied by Precise Enumerative Methods: Regular Periodical Increase of the Parasites Disclosed (with David Thomson) (1910)
- Discussion on the Treatment of Malaria (1918)
- Mosquitoes and Malaria in Britain (1918)
- Suggestions for the Care of Malaria Patients (1919)
- Observations on malaria (1919)

- Memoirs, with a Full Account of the Great Malaria Problem and Its Solution (1923)
- Malaria-control in Ceylon Plantations (1926)
- Solid Space-algebra: The Systems of Hamilton and Grassmann Combined (1929)
- A Summary of Facts Regarding Malaria Suitable for Public Instruction (with Malcolm Watson) (1930)
- Memories of Sir Patrick Manson (1930)
- The solution of equations by iteration (with William Stott) (1930)
- A Priori Pathometry (with H.P. Hudson) (1931)
- Mosquito Brigades and How to Organize Them (1902)

Annex 13. **Works of Ross on Parasites other than Malaria**

1. In the year 1903, Ross wrote a paper titled 'Note on the bodies recently described by Leishman and Donovan' (B.M.J. 1903, 2: 1261-1262).
2. Ross also wrote 'Further notes of Leishman's bodies' (B.M.J. 1903, 2: 1401).

Annex 14. **Autobiograhy by Dr Ronald Ross**

1. 1923. *Memoirs, with a Full Account of the Great Malaria Problem and its Solution.* London: John Murray. 547 pp.

Annex 15. **Places in India where Dr Ronald Ross Worked**

1. Madras (stationed in Madras and Bangalore in between 1881-1888)
2. Secunderabad- Deccan (1895; 1897)
3. Bangalore (1895-1897)
4. Ootacamund (1897)
5. Bombay (1897)
6. Oodeypore (1897)
7. Kherwara, Rajputana/Rajasthan (1897-1898)
8. Nasirabad (Rajasthan)
9. Calcutta (1898-1899)
10. Punkarbari, Darjeeling Dist. (1898)
11. Kurseong (1898)
12. Nowgong, Assam (1898)

Annex 16. **Annotated Catalogue of Extant Species in India**

(* species already described before the era of Dr Ronald Ross)

CATALOGUE OF INDIAN MOSQUITOES

Order: Insecta

Family: Culicidae

Subfamily: Anophelinae

Tribe: Anophelini

I . **Genus *ANOPHELES* Meigen, 1818**

(i) Subgenus *Anopheles* Meigen, 1818

1. *Anopheles (Anopheles) ahomi* Chowdhury, 1929
2. *Anopheles (Anopheles) aitkenii James,* 1903
3. *Anopheles (Anopheles) annandalei* Prasad, 1918
4. *Anopheles (Anopheles) argyropus* (Swellengrebel, 1914)
5. *Anopheles (Anopheles) baileyi* Edwards,1929
6. *Anopheles (Anopheles) barbirostris* van der Wulp, 1884
7. *Anopheles (Anopheles) barbumbrosus* Strickland & Chowdhury, 1927
8. *Anopheles (Anopheles) barianensis* James, 1911
9. *Anopheles (Anopheles) bengalensis* Puri, 1930
10. *Anopheles (Anopheles) crawfordi* Reid, 1953
11. *Anopheles (Anopheles) culiciformis* Cogill, 1903
12. *Anopheles (Anopheles) gigas* Giles, 1901
13. *Anopheles (Anopheles) hodgkini* Reid, 1962
14. *Anopheles (Anopheles) insulaeflorum* (Swellengrebel & Swellengrebel de Graaf, 1920)
15. *Anopheles (Anopheles) interruptus* Puri, 1929
16. *Anopheles (Anopheles) lesteri paraliae* Sandosham 1959
17. *Anopheles (Anopheles) lindesayi Giles*, 1900
18. *Anopheles (Anopheles) nigerrimus Giles*, 1900

19. *Anopheles (Anopheles) nilgiricus Christophers*, 1924
20. *Anopheles (Anopheles) nitidus* Harrison, Scanlon & Reid, 1973
21. *Anopheles (Anopheles) peditaeniatus* (Leicester, 1908)
22. *Anopheles (Anopheles) pinjaurensis* Barraud, 1932
23. *Anopheles (Anopheles) roperi* Reid, 1950
24. *Anopheles (Anopheles) sinensis* Wiedemann, 1828
25. *Anopheles (Anopheles) sintoni* Puri, 1929
26. *Anopheles (Anopheles) umbrosus* (Theobald, 1903)

(ii) Subgenus *Cellia* Theobald, 1902

27. *Anopheles (Cellia) aconitus* Donitz, 1902
28. *Anopheles (Cellia) annularis* van der Wulp, 1884
29. *Anopheles (Cellia) baimaii* Sallum & Peyton, 2005
30. *Anopheles (Cellia) balabacensis* Baisas, 1936
31. *Anopheles (Cellia) culicifacies* Giles, 1901
32. *Anopheles (Cellia) dravidicus* Christophers, 1924
33. *Anopheles (Cellia) dthali* Patton, 1905
34. *Anopheles (Cellia) elegans* (James, 1903)
35. *Anopheles (Cellia) fluviatilis* James, 1902
36. *Anopheles (Cellia) jamesii* Theobald, 1901
37. *Anopheles (Cellia) jeyporiensis* James, 1902
38. *Anopheles (Cellia) karwari* (James, 1903)
39. *Anopheles (Cellia) kochi* Donitz, 1901
40. *Anopheles (Cellia) maculatus* Theobald, 1901
41. *Anopheles (Cellia) majidi* Young & Majid, 1928
42. *Anopheles (Cellia) minimus* Theobald, 1901
43. *Anopheles (Cellia) mirans* Sallum & Peyton, 2005
44. *Anopheles (Cellia) moghulensis* Christophers, 1924
45. *Anopheles (Cellia) multicolor* Cambouliu, 1902
46. *Anopheles (Cellia) nivipes* (Theobald, 1903)
47. *Anopheles (Cellia) pallidus* Theobald, 1901
48. *Anopheles (Cellia) philippinensis* Ludlow, 1902
49. *Anopheles (Cellia) pseudojamesi* Strickland & Chowdhury, 1927
50. *Anopheles (Cellia) pseudosundaicus* Tyagi *et al.*, 2009
51. *Anopheles (Cellia) pseudowillmori* (Theobald, 1910)
52. *Anopheles (Cellia) pulcherrimus* Theobald, 1902

53. *Anopheles (Cellia) sergentii* (Theobald, 1907)
54. *Anopheles (Cellia) splendidus* Koidzumi, 1920
55. *Anopheles (Cellia) stephensi* Liston, 1901
56. *Anopheles (Cellia) subpictus* Grassi, 1899
57. *Anopheles (Cellia) sundaicus* (Rodenwaldt, 1925)
58. *Anopheles (Cellia) tessellatus* Theobald, 1901
59. *Anopheles (Cellia) theobaldi* Giles, 1901
60. *Anopheles (Cellia) turkhudi* Liston, 1901
61. *Anopheles (Cellia) vagus* Donitz, 1902
62. *Anopheles (Cellia) varuna* Iyengar, 1924
63. *Anopheles (Cellia) willmori* (James, 1903)

Subfamily: Culicinae

Tribe : Aedeomyiini Theobald, 1901

I. Genus *AEDEOMYIA* Theobald, 1901

(i) Subgenus Aedeomyia Theobald, 1901

64. *Aedeomyia (Aedeomyia) catasticta* Knab, 1909
65. *Aedeomyia (Aedeomyia) venustipes* (Skuse, 1889)

Tribe : Aedini Neveu-Lemaire, 1902

I. Genus *AEDIMORPHUS* Theobald, 1903

66. *Aedimorphus alboscutellatus* (Theobald, 1905)
67. *Aedimorphus caecus* (Theobald, 1901)
68. *Aedimorphus culicinus* (Edwards, 1922)
69. *Aedimorphus jamesi* (Edwards, 1914)
70. *Aedimorphus lowisii* (Theobald, 1910)
71. *Aedimorphus mediolineatus* (Theobald, 1901)
72. *Aedimorphus nigrostriatus* Barraud, 1927
73. *Aedimorphus pallidostriatus* (Theobald, 1907)
74. *Aedimorphus pampangensis* (Ludlow, 1905)
75. *Aedimorphus pipersalatus* (Giles, 1902)
76. *Aedimorphus punctifemoris* (Ludlow, 1921)
77. *Aedimorphus stenoetrus* (Theobald, 1907)
78. *Aedimorphus syntheticus* (Barraud, 1928)
79. *Aedimorphus taeniorhynchoides* (Christophers, 1911)

80. *Aedimorphus trimaculatus* (Theobald, 1905)
81. *Aedimorphus vexans* (Meigen, 1830)

II. Genus *ARMIGERES* Theobald, 1901

(i) Subgenus Armigeres Theobald,1901

82. *Armigeres (Armigeres) aureolineatus* (Leicester, 1908)
83. *Armigeres (Armigeres) durhami* (Edwards, 1917)
84. *Armigeres (Armigeres) joloensis* (Ludlow, 1904)
85. *Armigeres (Armigeres) kesseli* Ramalingam, 1987
86. *Armigeres (Armigeres) kuchingensis* Edwards, 1915
87. *Armigeres (Armigeres) mahantai* Bhattacharyya *et al.*, 2009
88. *Armigeres (Armigeres) pallithorax* Dong, Zhou & Dong, 2004
89. *Armigeres (Armigeres) subalbatus* (Coquillett, 1898)
90. *Armigeres (Armigeres) theobaldi* Barruad, 1934

(ii) Subgenus Leicesteria Theobald, 1904

91. *Armigeres (Leicesteria) annulipalpis* (Theobald, 1910)
92. *Armigeres (Leicesteria) annulitarsis* (Leicester, 1908)
93. *Armigeres (Leicesteria) dentatus* Barraud, 1927
94. *Armigeres (Leicesteria) digitatus* (Edwards, 1914)
95. *Armigeres (Leicesteria) dolichocephalus* (Leicester, 1908)
96. *Armigeres (Leicesteria) flavus* (Leicester, 1908)
97. *Armigeres (Leicesteria) inchoatus* Barraud, 1927
98. *Armigeres (Leicesteria) longipalpis* (Leicester, 1904)
99. *Armigeres (Leicesteria) magnus* (Theobald, 1908)
100. *Armigeres (Leicesteria) omissus* (Edwards, 1914)

III. Genus *AYURAKITIA* Thurman, 1954

101. *Ayurakitia peytoni* (Reinert, 1972)

IV. Genus *BRUCEHARRISONIUS* Reinert, 2003

102. *Bruceharrisonius aureostriatus* (Doleschall, 1857)
103. *Bruceharrisonius christophersi* (Edwards, 1922)
104. *Bruceharrisonius doonii* (Wattal, Bhatia & Kalra, 1958)
105. *Bruceharrisonius greenii* (Theobald, 1903)

V. Genus *CANCRAEDES* Edwards, 1929

106. *Cancraedes cancricomes* (Edwards, 1922)

107. *Cancraedes simplex* (Theobald, 1903)

VI. Genus *CHRISTOPHERSIOMYIA* Barraud, 1923

108. *Christophersiomyia annulirostris* (Theobald, 1905)
109. *Christophersiomyia gombakensis* (Mattingly, 1959)
110. *Christophersiomyia ibis* (Barraud, 1931)
111. *Christophersiomyia thomsoni* (Theobald, 1905)

VII. Genus *COLLESSIUS* Reinert, Harbach & Kitching, 2006

(i) Subgenus *Alloeomyia* Reinert, Harbach & Kitching, 2008

112. *Collessius (Alloeomyia) pseudotaeniatus* (Giles, 1901)

(ii) Subgenus *Collessius* Reinert, Harbach & Kitching, 2006

113. *Collessius (Collessius) elsiae* (Barraud, 1923)
114. *Collessius (Collessius) macdougalli* (Edwards, 1922)
115. *Collessius (Collessius) shortti* (Barraud, 1923)

VIII. Genus *DANIELSIA* Theobald, 1904

116. *Danielsia albotaeniata* Leicester, 1904
117. *Danielsia lepchana* (Barruad, 1923)

IX. Genus *DENDROSKUSEA* Edwards, 1929

118. *Dendroskusea kanarensis* (Edwards, 1934)
119. *Dendroskusea micropterus* (Giles, 1901)
120. *Dendroskusea periskelata* (Giles, 1902)
121. *Dendroskusea ramachandrai* (Reuben, 1967)
122. *Dendroskusea reginae* (Edwards, 1922)

X. Genus *DOWNSIOMYIA* Vargas, 1950

123. *Downsiomyia albolateralis* (Theobald, 1908)
124. *Downsiomyia albonivea* (Barruad,1934)
125. *Downsiomyia mohani* (Knight, 1969)
126. *Downsiomyia nivea* (Ludlow,1903)
127. *Downsiomyia niveoides* (Barruad, 1934)
128. *Downsiomyia novonivea* (Barraud,1934)

XI. Genus *EDWARDSAEDES* Belkin,1962

129. *Edwardsaedes imprimens* (Walkar, 1860)

XII. Genus *FINLAYA* Theobald, 1903

130. Finlaya flavipennis Giles, 1904
131. *Finlaya poicilia* Theobald, 1903

XIII. Genus *FREDWARDSIUS* Reinert, 2000

132. *Fredwardsius vittatus* (Bigot, 1861)

XIV. Genus *GILESIUS* Reinert, Harbach & Kitching, 2006

133. *Gilesius pulchriventer* (Giles, 1901)

XV. Genus *HEIZMANNIA* Ludlow, 1905

(i) Subgenus *Heizmannia* Ludlow, 1905

134. *Heizmannia (Heizmannia) aureochaeta* (Leicester, 1908)
135. *Heizmannia (Heizmannia) chandi* Edwards, 1922
136. *Heizmannia (Heizmannia) chengi* Lien, 1968
137. *Heizmannia (Heizmannia) complex* (Theobald, 1910)
138. *Heizmannia (Heizmannia) covelli* Barraud, 1929
139. *Heizmannia (Heizmannia) funerea* (Leicester, 1908)
140. *Heizmannia (Heizmannia) greenii* (Theobald, 1905)
141. *Heizmannia (Heizmannia) himalayensis* Edwards, 1922
142. *Heizmannia (Heizmannia) indic*a (Theobald, 1905)
143. *Heizmannia (Heizmannia) metallica* Leicester, 1908
144. *Heizmannia (Heizmannia) reidi* Mattingly, 1957
145. *Heizmannia (Heizmannia) viridis* Barraud, 1929

(ii) Subgenus *Mattinglyia* Lien, 1968

146. *Heizmannia (Mattinglyia) discrepans* (Edwards, 1922)
147. *Heizmannia (Mattinglyia) tripunctata* (Theobald, 1908)

XVI. Genus *HIMALAIUS* Reinert, Harbach & Kitching, 2006

148. *Himalaius gilli* (Barraud, 1924)
149. *Himalaius simlensis* (Edwards, 1922)

XVII. Genus *HOPKINSIUS* Reinert, Harbach & Kitching, 2008

(i) Subgenus *Yamada* Reinert, Harbach & Kitching, 2008

150. *Hopkinsius (Yamada) albocinctus* (Barraud, 1924)

XVIII. Genus *HULECOETEOMYIA* Theobald, 1904

151. *Hulecoeteomyia chrysolineata* (Theobald, 1907)

152. *Hulecoeteomyia formosensis* (Yamada, 1921)
153. *Hulecoeteomyia harveyi* (Barraud, 1923)
154. *Hulecoeteomyia pallirostris* (Edwards, 1922)
155. *Hulecoeteomyia saxicola* (Edwards, 1922)
156. *Hulecoeteomyia cherrapunjiensis* Natarajan *et al.*, 2016

XIX. Genus *INDUSIUS* Edwards, 1934

157. *Indusius pulverulentus* (Edwards, 1922)

XX. Genus *JIHLIENIUS* Reinert, Harbach & Kitching, 2006

158. *Jihlienius unicinctus* (Edwards, 1922)

XXI. Genus *KENKNIGHTIA* Reinert, 1990

159. *Kenknightia dissimilis* (Leicester, 1908)
160. *Kenknightia karwari* (Barraud, 1924)

XXII. Genus *LORRAINEA* Belkin,1962

161. *Lorrainea amesii* (Ludlow, 1903)
162. *Lorrainea fumida* (Edwards, 1928)

XXIII. Genus *MUCIDUS* Theobald, 1901

(i) Subgenus *Mucidus* Theobald, 1901

163. *Mucidus (Mucidus) laniger* (Wiedemann, 1820)
164. *Mucidus (Mucidus) quasiferinus* (Mattingly 1961)
165. *Mucidus (Mucidus) scatophagoides* Theobald, 1901

XXIV. Geunus *NEOMELANICONION* Newstead, 1907

166. *Neomelaniconion lineatopenne* (Ludlow, 1905)

XXV. Genus *OCHLEROTATUS* Lynch Arribalzaga, 1891 *Sensu*

(i) Subgenus *Finlaya* Theobald, 1903 *Sensu auctorum*

167. *Ochlerotatus (Finlaya) auronitens* (Edwards, 1922)
168. *Ochlerotatus (Finlaya) oreophilus* Edwards, 1916
169. *Ochlerotatus (Finlaya) sintoni* (Barraud, 1924)
170. *Ochlerotatus (Finlaya) suffusus* (Edwards, 1922)
171. *Ochlerotatus (Finlaya) versicolor* (Barraud,1924)

Subgenus uncertain

172. *Ochlerotatus pulcritarsis* (Rondani, 1872)

173. *Ochlerotatus pullatus* (Coquillett, 1904)

XXVI. Genus *PARAEDES* Edwards, 1934

174. *Paraedes barraudi* Edwards, 1934
175. *Paraedes chrysoscuta* (Theobald, 1910)
176. *Paraedes menoni* (Mattingly, 1958)
177. *Paraedes ostentatio* (Leicester, 1908)

XXVII. Genus *PETERMATTINGLYIUS* Reinert, Harbach & Kitching, 2009

(i) Subgenus *Petermattinglyius* Reinert, Harbach & Kitching, 2009

178. *Petermattinglyius (Petermattinglyius) franciscoi* (Mattingly, 1959)
179. *Petermattinglyius (Petermattinglyius) iyengari* (Edwards, 1923)

XXVIII. Genus *PHAGOMYIA* Theobald,1905

180. *Phagomyia assamensis* (Theobald 1908)
181. *Phagomyia cacharana* (Barraud, 1923)
182. *Phagomyia cogilli* (Edwards, 1922)
183. *Phagomyia deccana* (Barraud, 1923)
184. *Phagomyia feegradei* (Barraud, 1934)
185. Phagomyia gubernatoris (Giles, 1901)
186. *Phagomyia inquinata* (Edwards, 1922)
187. *Phagomyia khazani* (Edwards, 1922)
188. *Phagomyia lophoventralis* (Theobald, 1910)
189. *Phagomyia prominens* (Barraud, 1923)
190. *Phagomyia stevensoni* (Barraud, 1923)

XXIX. Genus *RHINOSKUSEA* Edwards, 1929

191. *Rhinoskusea longirostris* (Leicester, 1908)
192. *Rhinoskusea portonovoensis* (Tewari & Hiriyan, 1992)
193. *Rhinoskusea wardi* (Reinert, 1976)

XXX. Genus *SCUTOMYIA* Theobald, 1904

194. *Scutomyia albolineata Theobald*, 1904

XXXI. Genus *STEGOMYIA* Theobald, 1901

(i) Subgenus *Actinothrix* Reinert, Harbach & Kitching, 2009

195. *Stegomyia (Actinothrix) edwardsi* Barraud, 1923

196. *Stegomyia (Actinothrix) seampi* (Huang, 1974)

(ii) Subgenus *Heteraspidion* Reinert, Harbach & Kitching, 2009

197. *Stegomyia (Heteraspidion) annandalei* Theobald, 1910
198. *Stegomyia (Heteraspidion) craggi* Barraud, 1923

(iii) Subgenus *Huangmyia* Reinert, Harbach & Kitching, 2009

199. *Stegomyia (Huangmyia) mediopunctata* Theobald, 1905
200. *Stegomyia (Huangmyia) perplexa* Leicester, 1908

(iv) Subgenus *Stegomyia* Theobald, 1901

201. *Stegomyia (Stegomyia) aegypti* (Linnaeus, 1762)

(v) Subgenus *Xyele* Reinert, Harbach & Kitching, 2009

202. *Stegomyia (Xyele) desmotes* Giles, 1904
[*subgenus unassigned yet for following*]
203. *Stegomyia albopicta* (Skuse, 1895)
204. *Stegomyia flavopicta* (Yamada, 1921)
205. *Stegomyia gardnerii imitator* (Leicester, 1908)
206. *Stegomyia krombeini* (Huang, 1975)
207. *Stegomyia malayensis* (Colless, 1962)
208. Stegomyia novalbopicta (Barraud, 1931)
209. *Stegomyia patriciae* (Mattingly, 1954)
210. *Stegomyia pseudalbopicta* Borel, 1928
211. *Stegomyia scutellaris* (Walker, 1859)
212. *Stegomyia subalbopicta* (Barraud, 1931)
213. *Stegomyia unilineata* (Theobald, 1906)
214. *Stegomyia w-alba* Theobald, 1905

XXXII. Genus *TEWARIUS* Reinert, 2006

215. *Tewarius agastyai* (Tewari & Hiriyan, 1992)
216. *Tewarius nummatus* (Edwards, 1923)
217. *Tewarius reubenae* (Tewari & Hiriyan, 1992)

XXXIII. Genus *UDAYA* Thurman, 1954

218. *Udaya argyrurus* (Edwards, 1934)
219. *Udaya subsimilis* (Barraud, 1927)

XXXIV. Genus *VERRALLINA* Theobald, 1903

(i) Subgenus *Harbachius* Reinert, 1999

220. *Verrallina (Harbachius) abdita* (Barraud, 1931)

221. *Verrallina (Harbachius) consonensis* (Reinert, 1973)
222. *Verrallina (Harbachius) uniformis* (Theobald, 1910)
223. *Verrallina (Harbachius) yusafi* (Barraud, 1931)

(ii) Subgenus *Neomacleaya* Theobald,1907

224. *Verrallina (Neomacleaya) agrestis* (Barraud, 1931)
225. *Verrallina (Neomacleaya) andamanensis* (Edwards, 1922)
226. *Verrallina (Neomacleaya) assamensis* Bhattacharyya *et al.*, 2004
227. *Verrallina (Neomacleaya) atria* (Barraud, 1928)
228. *Verrallina (Neomacleaya) cauta* (Barraud, 1928)
229. *Verrallina (Neomacleaya) clavata* (Barraud, 1931)
230. *Verrallina (Neomacleaya) comata* (Barraud, 1931)
231. *Verrallina (Neomacleaya) hirsutipleura* Barraud, 1928
232. *Verrallina (Neomacleaya) indica* (Theobald, 1907)
233. *Verrallina (Neomacleaya) pseudodiurna* (Theobald, 1910)
234. *Verrallina (Neomacleaya) pseudomediofasciata* (Theobald, 1910)
235. *Verrallina (Neomacleaya) rami* (Barraud, 1928)
236. *Verrallina (Neomacleaya) seculata* (Menon, 1950)
237. *Verrallina (Neomacleaya) vallistris* (Barraud, 1928)
238. *Verrallina (Neomacleaya) unca* (Theobald,1901)

(iii) Subgenus *Verrallina* Theobald,1903

239. *Verrallina (Verrallina) butleri* (Theobald, 1901)
240. *Verrallina (Verrallina) dux* (Dyar & Shannon, 1925)
241. *Verrallina (Verrallina) lugubris* (Barraud, 1928)

Tribe: Culicini Meigen, 1818

I. Genus *CULEX* Linnaeus, 1758

(i) Subgenus *Barraudius* Edwards, 1921

242. *Culex (Barraudius) modestus* Ficalbi, 1890

(ii) Subgenus *Culex* Linnaeus, 1758

243. *Culex (Culex) alienus* Colless, 1957
244. *Culex (Culex) annulus* Theobald, 1901
245. *Culex (Culex) barraudi* Edwards, 1922
246. *Culex (Culex) edwardsi* Barraud, 1923
247. *Culex (Culex) fuscifurcatus* Edwards, 1937
248. *Culex (Culex) fuscocephala* Theobald, 1907

249. *Culex (Culex) gelidus* Theobald, 1901
250. *Culex (Culex) hutchinsoni* Barraud, 1924
251. *Culex (Culex) jacksoni* Edwards, 1934
252. *Culex (Culex) mimeticus* Noe, 1899
253. *Culex (Culex) mimuloides* Barraud, 1924
254. *Culex (Culex) mimulus* Edwards, 1915
255. *Culex (Culex) murrelli* Lien, 1968
256. *Culex (Culex) nilgiricus* Edwards, 1916
257. *Culex (Culex) perexiguus* Theobald, 1903
258. *Culex (Culex) perplexus* Leicester, 1908
259. *Culex (Culex) pseudovishnui* Colless, 1957
260. *Culex (Culex) quinquefasciatus* Say, 1823
261. *Culex (Culex) sitiens* Wiedemann, 1828
262. *Culex (Culex) theileri* Theobald, 1903
263. *Culex (Culex) tritaeniorhynchus* Giles, 1901
264. *Culex (Culex) univittatus* Theobald, 1901
265. *Culex (Culex) vagans* Wiedemann, 1828
266. *Culex (Culex) vishnui* Theobald, 1901
267. *Culex (Culex) whitei* Barraud, 1923
268. *Culex (Culex) whitmorei* (Giles, 1904)

(iii) Subgenus *Culiciomyia* Theobald, 1907

269. *Culex (Culiciomyia) bailyi* Barraud, 1934
270. *Culex (Culiciomyia) fragilis* Ludlow, 1903
271. *Culex (Culiciomyia) harrisoni* Sirivanakarn, 1977
272. *Culex (Culiciomyia) nigropunctatus* Edwards, 1926
273. *Culex (Culiciomyia) pallidothorax* Theobald, 1905
274. *Culex (Culiciomyia) ramakrishnii* Wattal & Kalra, 1965
275. *Culex (Culiciomyia) scanloni* Bram, 1967
276. *Culex (Culiciomyia) shebbearei* Barraud, 1924
277. *Culex (Culiciomyia) spathifurca* (Edwards, 1915)
278. *Culex (Culiciomyia) viridiventer* Giles, 1901

(iv) Subgenus *Eumelanomyia* Theobald, 1909

279. *Culex (Eumelanomyia) brevipalpis* (Giles, 1902)
280. *Culex (Eumelanomyia) castrensis* Edwards, 1922
281. *Culex (Eumelanomyia) foliatus* Brug, 1932

282. *Culex (Eumelanomyia) hinglungensis* Chu, 1957
283. *Culex (Eumelanomyia) iphis* Barraud, 1924
284. *Culex (Eumelanomyia) khazani* Edwards, 1922
285. *Culex (Eumelanomyia) malayi* (Leicester, 1908)
286. *Culex (Eumelanomyia) mohani* Sirivanakarn, 1977
287. *Culex (Eumelanomyia) pluvialis* Barraud, 1924
288. *Culex (Eumelanomyia) tenuipalpis* Barraud, 1924

(v) Subgenus *Lophoceraomyia* Theobald, 1905

289. *Culex (Lophoceraomyia) aculeatus* Colless, 1965
290. *Culex (Lophoceraomyia) bengalensis* Barraud, 1934
291. *Culex (Lophoceraomyia) bicornutus* (Theobald, 1910)
292. *Culex (Lophoceraomyia) cinctellus* Edwards, 1922
293. *Culex (Lophoceratomyia) cubitatus* Colless, 1965
294. *Culex (Lophoceraomyia) demissus* Colless, 1965
295. *Culex (Lophoceraomyia) flavicornis* Barraud, 1924
296. *Culex (Lophoceraomyia) gracicornis* Sirivanakarn, 1977
297. *Culex (Lophoceraomyia) inculus* Colless, 1965
298. *Culex (Lophoceraomyia) infantulus* Edwards, 1922
299. *Culex (Lophoceraomyia) lasiopalpis* Sirivanakarn, 1977
300. *Culex (Lophoceraomyia) macdonaldi* Colless, 1965
301. *Culex (Lophoceraomyia) mammilifer* (Leicester, 1908)
302. *Culex (Lophoceraomyia) minor* (Leicester, 1908)
303. *Culex (Lophoceraomyia) minutissimus* (Theobald, 1907)
304. *Culex (Lophoceraomyia) paraculeatus* Sirivanakarn, 1977
305. *Culex (Lophoceraomyia) peytoni* Bram & Rattanarithikul, 1967
306. *Culex (Lophoceraomyia) pholeter* Bram & Rattanarithikul, 1967
307. *Culex (Lophoceraomyia) pilifemoralis* Wang & Feng, 1964
308. *Culex (Lophoceraomyia) quadripalpis* (Edwards, 1914)
309. *Culex (Lophoceraomyia) raghavanii* Rahman, Chowdhury & Kalra, 1968
310. Culex (Lophoceraomyia) *rubithoracis* (Leicester, 1908)
311. Culex (Lophoceraomyia) *seniori Barraud*, 1934
312. Culex (Lophoceraomyia) *singhbhumensis* Natarajan & Rajavel, 2009
313. Culex (Lophoceraomyia) *uniformis* (Theobald, 1905)
314. Culex (Lophoceraomyia) *variatus* (Leicester, 1908)
315. Culex (Lophoceraomyia) *wilfredi* Colless, 1965

(vi) Subgenus *Maillotia* Theobald, 1907

316. *Culex (Maillotia) hortensis* Ficalbi, 1889

(vii) Subgenus *Oculeomyia* Theobald, 1907

317. *Culex (Oculeomyia) bitaeniorhynchus* Giles, 1901
318. *Culex* (*Oculeomyia*) *cornutus* Edwards, 1922
319. *Culex* (*Oculeomyia*) *epidesmus* (Theobald, 1910)
320. *Culex* (*Oculeomyia*) *infula* Theobald, 1901
321. *Culex* (*Oculeomyia*) *luzonensis* Sirivanakarn, 1976
322. *Culex* (*Oculeomyia*) *sinensis* Theobald, 1903

II. Genus *LUTZIA* Theobald, 1903

(i) Subgenus *Metalutzia* Tanaka, 2003

323. *Lutzia* (*Metalutzia*) *agranensis* Singh & Prakash, 2008
324. *Lutzia* (*Metalutzia*) *fuscana* (Wiedemann, 1820)
325. *Lutzia (Metalutzia) halifaxii* (Theobald, 1903)
326. *Lutzia (Metalutzia) vorax* Edwards, 1921

Tribe: Culisetini Belkin, 1962

I. Genus *CULISETA* Felt, 1904

(i) Subgenus *Allotheobaldia* Broelemann,1919

327. *Culiseta (Allotheobaldia) longiareolata* (Macquart, 1838)

(ii) Subgenus *Culiseta* Felt,1904

328. *Culiseta (Culiseta) alaskaensis indica* (Edwards, 1920)
329. *Culiseta (Culiseta) niveitaeniata* (Theobald, 1907)

Tribe: Ficalbiini Belkin, 1962

I. Genus *FICALBIA* Theobald, 1903

330. *Ficalbia minima* (Theobald, 1901)

II. Genus *MIMOMYIA* Theobald, 1903

(i) Subgenus *Etorleptiomyia* Theobald, 1904

331. *Mimomyia (Etorleptiomyia) luzonensis* (Ludlow, 1905)

(ii) Subgenus *Ingramia* Edwards, 1912

332. *Mimomyia* (*Ingramia*) *fusca* (Leicester, 1908)

(iii) Subgenus *Mimomyia* Theobald, 1903

333. *Mimomyia* (*Mimomyia*) *aurea* (Leicester, 1908)

334. *Mimomyia (Mimomyia) chamberlaini* Ludlow, 1904
ssp. *clavipalpus* (Theobald, 1908)
335. *Mimomyia* (*Mimomyia*) *hybrida* (Leicester, 1908)
336. *Mimomyia (Mimomyia) intermedia* (Barraud, 1929)

Tribe: Hodgesiini Belkin,1962

I. Genus *HODGESIA* Theobald, 1904

337. *Hodgesia bailyi* Barraud, 1929
338. *Hodgesia malayi* Leicester, 1908

Tribe : Mansoniini Belkin,1962

I. Genus *COQUILLETTIDIA* Dyar, 1905

(i) Subgenus *Coquillettidia* Dyar, 1905

339. *Coquillettidia (Coquillettidia) crassipes* (van der Wulp, 1881)
340. *Coquillettidia (Coquillettidia) novochracea* (Barraud, 1927)
341. *Coquillettidia (Coquillettidia) ochracea* (Theobald, 1903)

II. Genus *MANSONIA* Blanchard, 1901

(i) Subgenus *Mansonioides* Theobald, 1907

342. *Mansonia (Mansonioides) annulifera* (Theobald, 1901)
343. *Mansonia* (*Mansonioides*) *dives* (Schiner, 1868)
344. *Mansonia (Mansonioides) indiana* Edwards, 1930
345. *Mansonia (Mansonioides) uniformis* (Theobald, 1901)

Tribe: Orthopodomyiini Belkin, Heinemann & Page, 1970

I. Genus *ORTHOPODOMYIA* Theobald, 1904

346. *Orthopodomyia albipes* Leicester, 1904
347. *Orthopodomyia andamanensis* Barraud, 1934
348. *Orthopodomyia anopheloides* (Giles, 1903)
349. *Orthopodomyia flavicosta* Barraud, 1927
350. *Orthopodomyia flavithorax* Barraud, 1927

Tribe: Sabethini Blanchard,1905

I. Genus *MALAYA* Leicester, 1908

351. *Malaya genurostris* Leicester, 1908
352. *Malaya jacobsoni* (Edwards, 1930)

II. Genus *TOPOMYIA* Leicester, 1908

(i) Subgenus *Suaymyia* Thurman, 1959

353. *Topomyia (Suaymyia) cristata* Thurman,1959

(ii) Subgenus *Topomyia* Leicester, 1908

354. *Topomyia (Topomyia) aureoventer* (Theobald, 1910)
355. *Topomyia (Topomyia) bifurcata* Dong,Wang & Lu,1995
356. *Topomyia (Topomyia) hirtusa* Gong,1989

III. Genus *TRIPTEROIDES* Giles, 1904

(i) Subgenus *Rachionotomyia* Theobald, 1905

357. *Tripteroides (Rachionotomyia) affinis* (Edwards, 1913)
358. *Tripteroides (Rachionotomyia) aranoides* (Theobald, 1901)
359. *Tripteroides (Rachionotomyia) ceylonensis* (Theobald 1905)
360. *Tripteroides (Rachionotomyia) coonorensis* Mattingly,1981
361. *Tripteroides (Rachionotomyia) edwardsi* (Barraud, 1929)
362. *Tripteroides (Rachionotomyia) serratus* (Barraud, 1929)

(ii) Subgenus *Tripteroides* Giles, 1904

363. *Tripteroides (Tripteroides) indicus* (Barraud, 1929)
364. *Tripteroides (Tripteroides) powelli* (Ludlow,1909)
365. *Tripteroides (Tripteroides) similis* (Leicester, 1908)
366. *Tripteroides (Tripteroides) tarsalis* Delfinado & Hodges, 1968

Tribe: Toxorhynchitini Lahille, 1904

I. Genus *TOXORHYNCHITES* Theobald, 1901

(i) Subgenus *Toxorhynchites* Theobald, 1901

367. *Toxorhynchites (Toxorhynchites) albipes* (Edwards, 1922)
368. *Toxorhynchites (Toxorhynchites) edwardsi* (Barraud, 1924)
369. *Toxorhynchites (Toxorhynchites) gravely* (Edwards, 1921)
370. *Toxorhynchites (Toxorhynchites) kempi* (Edwards,1921)
371. *Toxorhynchites (Toxorhynchites) klossi* (Edwards, 1921)
372. *Toxorhynchites (Toxorhynchites) metallicus* Leicester 1904
373. *Toxorhynchites (Toxorhynchites) minimus* (Theobald, 1905)
374. *Toxorhynchites (Toxorhynchites) splendens* (Wiedemann, 1819)
375. *Toxorhynchites (Toxorhynchites) tyagii* Krishnamoorthy *et al.*, 2013

Tribe: Uranotaeniini Lahille,1904

I. Genus *URANOTAENIA* Lynch Arribalzaga, 1891

(i) Subgenus *Pseudoficalbia* Theobald, 1912

376. *Uranotaenia (Pseudoficalbia) atra* Theobald, 1905
377. *Uranotaenia (Pseudoficalbia) bicolor* Leicester, 1908
378. *Uranotaenia (Pseudoficalbia) bimaculata* Leicester, 1908
379. *Uranotaenia (Pseudoficalbia) dibrugarhensis* Bhattacharyya *et al.*, 2004
380. *Uranotaenia (Pseudoficalbia) husaini* Qutubuddin, 1947
381. *Uranotaenia (Pseudoficalbia) luteola* Edwards, 1934
382. *Uranotaenia (Pseudoficalbia) lutescens* Leicester, 1908
383. *Uranotaenia (Pseudoficalbia) maculipleura* Leicester, 1908
384. *Uranotaenia (Pseudoficalbia) mattinglyi* Qutubuddin, 1951
385. *Uranotaenia (Pseudoficalbia) maxima* Leicester, 1908
386. *Uranotaenia (Pseudoficalbia) nivipleura* Leicester, 1908
387. *Uranotaenia (Pseudoficalbia) novobscura* Barraud, 1934
388. *Uranotaenia (Pseudoficalbia) obscura* Edwards, 1915
389. *Uranotaenia (Pseudoficalbia) ohamai* Tanaka, Mizusawa & Saugstad, 1975
390. *Uranotaenia (Pseudoficalbia) recondita* Edwards, 1922
391. *Uranotaenia (Pseudoficalbia) stricklandi* Barraud, 1926
392. *Uranotaenia (Pseudoficalbia) unguiculata* Edwards, 1913
393. *Uranotaenia (Pseudoficalbia) maikalensis* Natarajan *et al.*, 2017
394. *Uranotaenia (Pseudoficalbia) dandakaranyensis* Natarajan *et al.*, 2017
395. *Uranotaenia (Pseudoficalbia) satpuraensis* Natarajan *et al.*, 2017

(ii) Subgenus *Uranotaenia* Lynch Arribalzaga, 1891

396. *Uranotaenia (Uranotaenia) alboannulata* (Theobald, 1905)
397. *Uranotaenia (Uranotaenia) annandalei* Barraud, 1926
398. *Uranotaenia (Uranotaenia) campestris* Leicester, 1908
399. *Uranotaenia (Uranotaenia) christophersi* Barraud, 1926
400. *Uranotaenia (Uranotaenia) edwardsi* Barraud, 1926
401. *Uranotaenia (Uranotaenia) hebes* Barraud, 1931
402. *Uranotaenia (Uranotaenia) lateralis* Ludlow, 1905
403. *Uranotaenia (Uranotaenia) longirostris* Leicester, 1908
404. *Uranotaenia (Uranotaenia) macfarlanei* Edwards, 1914
405. *Uranotaenia (Uranotaenia) micans* Leicester, 1908
406. *Uranotaenia (Uranotaenia) orientalis* Barraud, 1926
407. *Uranotaenia (Uranotaenia) testacea* Theobald, 1905

■